P9-CCR-028

THE 125 BEST
BRAIN TEASERS
OF ALL TIME

The
125
BEST
BRAIN
TEASERS
of All Time

A Mind-Blowing Challenge of
MATH, LOGIC, AND WORDPLAY

MARCEL DANESI, Ph.D.

**ZEPHYROS
PRESS**

For general information on our other products and services or to obtain technical support, please contact our Customer Care Department within the United States at (866) 744-2665, or outside the United States at (510) 253-0500.

Zephyros Press publishes its books in a variety of electronic and print formats. Some content that appears in print may not be available in electronic books, and vice versa.

TRADEMARKS: Zephyros Press and the Zephyros Press logo are trademarks or registered trademarks of Callisto Media Inc. and/or its affiliates, in the United States and other countries, and may not be used without written permission. All other trademarks are the property of their respective owners. Zephyros Press is not associated with any product or vendor mentioned in this book.

Illustration © Lan Truong, 2018 (cover); Drekhann/iStock.com (cover background and interior).

ISBN: Print 978-1-64152-008-9 | eBook 978-1-64152-009-6

R1

I dedicate this book to my three grandchildren, Alexander, Sarah, and Charlotte. They have solved my own puzzle of existence simply by being born and being so beautiful.

CONTENTS

✳

The Time-Honored Fun of
BRAIN TEASERS

*

SINCE THE DAWN OF CIVILIZATION, we have been fascinated by conundrums, rebuses, riddles, and enigmas of all kinds. The historical record has made this quite obvious, demonstrating our innate propensity for puzzles and games that has no parallel in any other species. The oldest known cipher—a message laid out in secret code—is a Sumerian text written in cuneiform (wedge-shaped markings carved in soft clay tablets), which dates back to around 2500 BCE. This text is one of the first examples of cryptography, now a popular genre of puzzles. Similar kinds of puzzles and games from the Old Babylonian period (1800–1600 BCE), Egypt (1700–1650 BCE), and the ancient civilizations of the Orient and the Americas have also been discovered by archaeologists. Even more ancient discoveries include game sticks in Africa going back ten thousand years.

One of the oldest puzzles known is the so-called Riddle of the Sphinx. In Greek mythology, the Sphinx was a monster with the head and breasts of a woman, the body of a lion, and the wings of a bird. Lying crouched on a rock, it stopped all those about to enter the city of Thebes by asking them a riddle. (You will find that riddle here in this book as puzzle 25.) Those who failed to answer the riddle correctly were killed on the spot. On the other hand, the Sphinx vowed to destroy itself if anyone managed to come up with the correct answer. When the hero Oedipus solved the riddle, the Sphinx killed itself as forewarned. For ridding them of this terrible monster, the Thebans crowned Oedipus their king.

Throughout history, puzzles have captivated the fancy of many famous personages. Riddle contests were organized by the biblical kings Solomon and Hiram. Charlemagne (742–814), the founder of the Holy Roman Empire, hired a scholar to create puzzles for various reasons. Edgar Allan Poe (1809–1849), the great American writer, and Lewis Carroll (1832–1898), best known for his two great children's novels, *Alice's Adventures in Wonderland* (1865) and *Through the Looking-Glass* (1872), made ingenious puzzles. The love of puzzles is ancient and shared by everyone, as witnessed today by the widespread popularity of puzzle magazines, "brain challenging" sections in newspapers, riddle books for children, TV quiz shows, websites, social media sites, and game tournaments. Millions of people the world over simply enjoy solving puzzles for their own sake. As the great British puzzlist Henry E. Dudeney (1857–1930) aptly put it, "A good puzzle, like virtue, is its own reward."

15 FASCINATING FACTS

1. There is no culture on earth without puzzle and game traditions. This strongly suggests that puzzles are hardwired in the brain and may serve some basic function in our species.

2. Many puzzles are associated with myth and legend. For example, arranging the first nine integers in a square pattern so that the sum of the numbers in each row, column, and diagonal is the same is called *Lo Shu* in China. (In English, it's Magic Square.) *Lo Shu* was invented 4,000 years ago, and the Chinese have always believed it

possesses mystical properties. To this day, it is thought to protect against the evil eye when placed over the entrance to a dwelling or room. I have one on the door to my office, and I carry another in my wallet, just in case!

3. Author Mark Twain (1835–1910) had this take on riddles (both of the puzzling and the philosophical variety): "Let us consider that we are all partially insane. It will explain us to each other; it will unriddle many riddles; it will make clear and simple many things which are involved in haunting and harassing difficulties and obscurities now."

4. The Alzheimer's Association in the United States has endorsed sudoku as a preventive therapy against the disease. Their recommendation is based on published studies. Other puzzle genres such as crosswords and jigsaw puzzles are also recommended. In other words, puzzles may indeed be therapy for the aging brain.

5. The number of possible sudoku puzzles that can be made with the first nine digits is calculated to be 6,670,903,752,021,072,936,960! It would take a computer over 211 billion years to solve them all.

6. The late puzzlist James Fixx (1932–1984) wrote, "Puzzles not only bring us pleasure, but also help us to work and learn more effectively." Consider that to be a general principle.

7. Cryptograms greatly appeal to our sense of mystery, as evidenced by their frequent use today in mystery and adventure movies, from *Sneakers* (1992) to *The Da Vinci Code* (2006) and *The Imitation Game* (2014).

8. Little is known of the Greek mathematician Metrodorus, who may have lived in the sixth century. His collection of

puzzles, which he called "epigrams" in his book the *Greek Anthology*, is still a challenging one, and rather contemporary in its style and content.

9. Archimedes (287–212 BCE) invented a game called the *loculus*, which was essentially a type of geometrical jigsaw puzzle. The objective was to combine the pieces in such a way that they fit seamlessly together.

10. Perhaps more than any other puzzle, the riddle genre has appeared in countless movies, including those as divergent as *Die Hard with a Vengeance* (1995) and *The Hobbit: An Unexpected Journey* (2012). The reason may well be that a riddle is effective for achieving multiple purposes, from keeping audiences in suspense to making indirect commentary on the plot.

11. Lateral-thinking puzzles truly sharpen the brain. Here's a classic one: How can it be that someone can fall from the window of a 50-story building and still survive? The answer: He or she fell from the ground-floor window.

12. The word "puzzle" was first documented in a book titled *The Voyage of Robert Dudley to the West Indies, 1594–1595* (1899) by Sir George F. Warner (1845–1936). It marked the first time the word was used to describe a type of game.

13. One of longest-running shows on TV is *Wheel of Fortune*, attesting to the popularity of word puzzles across generations.

14. Chess is an important element in *Blade Runner* (1982), suggesting that the game is a gauge of human creativity and intelligence. It begs the question: If an advanced robot can play chess, does that make it intelligent?

15. Perhaps the most famous use of chess in film is Ingmar Bergman's classic from 1957, *The Seventh Seal*. The scene in which a medieval knight challenges Death to a chess game to save his own life is one of the most memorable in the history of cinema.

HOW TO USE THIS BOOK

I love puzzles. I love researching what goes on in the mind as we solve them. The best brain teasers of all time are—as you might imagine—those that have withstood the test of time. In this book, you will find a number of "classic nuts," as they are called, in many cases paraphrased for convenience. Having written so many puzzles and taught on the subject, I believe that the classic ones I've selected here are among the best of all time. Many have their origins in distant centuries and far-flung cultures, but they are all united by being at once fun and challenging to solve. Although I have organized them from easy to challenging, the sequencing is bound to be subjective. One principle has guided me throughout—all puzzles should be "doable." Some may take more patience to crack, but they too should be understandable and eventually lead to a clear solution.

This book is designed to provide hours of entertainment and mental challenge. It contains 125 puzzles spread out over five levels of difficulty. The puzzles are numbered consecutively across the levels. Unless you are already an experienced puzzle-solver, you should go through them in order; otherwise skip a tough nut, and come back to it later. Clues are provided to help you for all puzzles—if you need them, of course.

Answers as well as step-by-step solutions for all the puzzles are given at the back. But you should not read these until you have attempted to solve the puzzles on your own first, no matter how frustrated you may be with any particular puzzle. After a while, you will get the knack of how to go about attacking the different genres of puzzles. If you discover the answer using a different line of attack than the one suggested in the book, you should still study the given solution, simply to get a different perspective on the puzzle-solving process itself. This, too, will enhance your puzzle-solving skills. Incidentally, these skills are not correlated with quickness of thought or IQ. A slow thinker can solve a puzzle just as successfully as a fast one can.

Puzzles can be both pleasurable and frustrating—they are fun if you solve them easily with an aha moment; they can be frustrating if you do not. Hopefully, you will achieve more of the former with this book. It offers intriguing and complex brain teasers (not cheap "gotcha" questions) from around the world and throughout history. It also provides puzzles in a variety of different types (math, logic, and wordplay), organized from easy questions to harder ones, turning the book into a progressively more challenging journey. It is designed to delight you by presenting mental challenges that are "doable" while making no assumptions.

If you are an inveterate puzzle-solver, you will recognize some of the classic ones here, paraphrased in specific ways. But there is ample material to keep you entertained, as well. If you only dabble in puzzles, consider this both a collection of classic puzzles with historical anecdotes and a set of original puzzles derived from them.

No specialized knowledge or intellectual tools are necessary to solve any of the brain teasers. To help you pace yourself, the difficulty level within each section is, more or less, uniform. But this can be hard to maintain or even determine, since puzzle-solving is, by its nature, very subjective. You may find solving Carroll's doublet puzzles or their derivatives easy but even a simple math puzzle very challenging.

So, there is really no single, "right" way to move through this book, although I suggest that you do start at the beginning if you are a beginner. Of course, you can always decide to skip around or plow straight through. The only requirement is that you challenge yourself (and your friends) and have a little fun.

10 PUZZLE-SOLVING TIPS

1. Before attempting to solve any specific puzzle, ask yourself the following questions: What does it ask me to do? What linguistic, logical, or mathematical feature is applicable? You have to grasp the basis of the puzzle first, before attempting to solve it.

2. If a puzzle seems particularly difficult or complicated, ask yourself: Can the puzzle be reduced to, or compared to, a simpler version that I have solved before?

3. Another way to approach a challenging puzzle is to ask yourself: What systematic approach (diagram, chart) can I employ, or what resources can I consult (dictionary, previous solutions)?

4. It is always useful to jot down the facts of the puzzle in a systematic and orderly fashion to get a solid grasp of all the essential elements.

5. To increase your chances of solving a tough puzzle, paraphrase it in your own words, inspecting and comparing the original to your version very carefully. You might get the relevant insight from this.

6. If you get frustrated over a puzzle, put it aside for a while, and come back to it at a later time. The unconscious mind will continue to work on it, even as you occupy yourself with something else. The solution may then come to you by a flash of insight.

7. Don't underestimate the usefulness of trial and error. If one approach works, end of story. If it does not, examine it to find out why it was fruitless. On that basis, try out a new hunch.

8. Always check your answer against the puzzle statement to verify its consistency with the original text.

9. If you do not get the correct solution to a puzzle, read the answer at the back of the book, which generally explains the reasoning behind the solution (unless it is a straightforward riddle, anagram, or the like). You might be able to extract general principles from it, which you can then apply to other puzzles.

10. Above all else, enjoy the simple pleasure of delving into a puzzle, even if it does not lead to a solution. Puzzles are miniature works of intellectual art and can be enjoyed as such. Lewis Carroll, who is probably the greatest puzzle-maker of all time, wrote puzzles in such a way that they can be read for their own sake.

Level

1

SMARTY-PANTS

1

Change-a-Letter: Saliva

The number of word games, many of them dating back to antiquity, is truly mind-boggling—riddles, anagrams, acrostics, and so on. So, let's start off with a classic type of word puzzle that involves letter changing, an offshoot of several forms of wordplay. For example, by changing one letter in a five-letter word meaning "frighten," you will get a word meaning "gaze" or "gape." One arrives at the solution by changing the *c* in *scare* ("frighten") to *t* to produce the word *stare* ("gaze" or "gape"). That's all there is to it.

> **Change one letter in a four-letter word that refers to *stain* to get a colloquial word for *saliva*.**

As you might expect, crosswords are among the most loved of all word puzzles. The inventor of the crossword puzzle was the British-born inventor Arthur Wynne (1871–1945). He created the first puzzle for the "Fun" section of the December 21, 1913, edition of the *New York World* and titled it "Word-Cross." However, due to a typesetting error, it appeared in print as "Cross-Word." That erroneous name stuck, and Wynne's puzzle has been known as *crossword* ever since.

Anagram: Admirers

Anagrams are words or phrases created by rearranging the letters of other words or phrases. Their origin goes right back to the dawn of recorded history. In antiquity, however, these word games were hardly perceived as mere entertainment; rather, they were believed to harbor secrets or prophetic messages. Below is a classic nut in this genre, meant for enjoyment, not divination!

If you rearrange the letters of the word *admirer* you will get a word referring to two people in a legal union. What word is that?

Anagrams have been a favorite pastime of many rulers and aristocrats of the past. For example, Louis XIII (1601–1643), the king of France during the early 17th century, hired a royal anagrammatist for a handsome salary. It was his job both to entertain the king and to create and send secret messages to the king's friends, either to stump them or provide mental recreation.

Coins, Coins

Puzzles and games based on the use of coins reach back considerably in time. The ancient Japanese called one of their most popular coin games *Hiroimono*, or "things picked up," because it is played by picking up and moving items one at a time. Our puzzle involves the value of coins, rather than moving them around in some way.

What three current US coins add up to 45 cents if one of the coins is not a nickel?

CLUE: PAGE 137 · ANSWER: PAGE 149

Dudeney's Relations

The following puzzle is based on a genre invented by one of the greatest puzzle-makers of all time, the Englishman Henry E. Dudeney. You will have an opportunity to solve his original puzzle—considerably more difficult than the version presented here—later on in this book.

A young girl answers her cell and asks, "Who is this?" A man's voice responds, "Figure it out. Your mom's mother is my mother-in-law." What's the man's relation to the girl?

Physicist Albert Einstein (1879–1955) once said, rather appropriately, that "Logic will get you from A to B. Imagination will take you everywhere."

Jobs, Jobs

Here is another of Dudeney's inventions paraphrased for convenience. If this is your first exposure to this kind of puzzle, you are in for a treat in logical thinking. By the way, here is what Dudeney had to say about puzzles: "The fact is that our lives are largely spent in solving puzzles; for what is a puzzle but a perplexing question? And from our childhood upwards we are perpetually asking questions or trying to answer them." Quite true, isn't it?

In a certain company, the positions of director, engineer, and accountant are held by Bob, Janet, and Shirley, but not necessarily in that order. The accountant, who is an only child, earns the least. Shirley, who is married to Bob's brother, earns more than the engineer. What position does each person fill?

As this puzzle demonstrates, logic and language are intertwined—indeed, the hidden logic is enshrouded in the implications of the words chosen. The famous Swiss psychologist Jean Piaget (1896–1980) put it aptly: "Logic and mathematics are nothing but specialized linguistic structures."

CLUE: PAGE 137 · ANSWER: PAGE 149

Compound Words

A compound word is the combination of two individual words to create an entirely new meaning. For example, the noun *handkerchief* is made up of two nouns: *hand* + *kerchief*. So, a handkerchief is a kerchief held in one's hand. Puzzles that play on this feature involve figuring out the compound word by relating the definition of its two parts. For example, can you determine the compound word made up of two distinct parts meaning "inside" and "place something?" The answer is *input*, since *in* = "inside" and *put* = "place something." It's as simple as that. Here are a few puzzles of this kind to challenge your brain.

What compound words do these suggest?

(1) companion + sailing vessel, (2) volume + retail establishment, (3) leaf color + abode, (4) gaze + away

The 1511 collection titled *Amusing Questions* was published by a printer called, rather appropriately, Wynkyn de Worde. It was among the first books printed after the invention of the printing press.

Racing: Three Runners

This form of logic puzzle is a real brain teaser. No one knows for sure who invented it, but it has been around for quite some time. This is the first of several such puzzles in this book, each successive example increasing in level of difficulty.

Three runners—Frieda, Hannah, and Gina—competed in a high school race. Frieda beat Hannah, but not Gina. Which runner came in first?

For some reason people have always been attracted to racing games, typically played with objects on surfaces and boards. Today, many of the most popular board games fall into this category, such as Parcheesi and backgammon.

CLUE: PAGE 138 · ANSWER: PAGE 150

Odd-One-Out: Materials

If you have ever taken an IQ test, you are familiar with the "odd-one-out" format. For example, in the set of words *baseball, tennis, helmet, football, hockey*, the word *helmet*, unlike the other four words, does not refer to a specific sport. Therefore, logically, it is the odd one out. While these puzzles commonly appear on IQ tests and are used to assess candidates for various purposes, surprisingly, nothing is known about their origin.

Which one of the words in this set does not belong conceptually to the others?

sand
iron
gasoline
copper
steel

Lateral Thinking: Seven Letters

You've heard of the expression *lateral thinking*, haven't you? It comes from psychologist Edward de Bono's work in the late 1960s, showing that sometimes we have to think outside the box a bit, as the expression goes. Since then, the lateral-thinking puzzle has become a genre all its own. One subgenre involves decoding the meaning of seemingly random symbols. This is the first of three classified here under the category of lateral thinking—with just a touch of poetic license.

What could the following sequence of letters possibly mean?

MTWTFSS

The classic example of lateral thinking is a story about a truck stuck under a low bridge. As a group of people tried to think of some way to force the truck out, a little boy, using lateral thinking, suggested a solution that is obvious, believe it or not. This puzzle will be presented to you later as number 57.

CLUE: PAGE 138 · ANSWER: PAGE 150

Coffee Math

Puzzles can be found in one of the first math textbooks ever devised, known as the *Ahmes Papyrus* from Egypt (1650 BCE). Throughout this puzzle collection, we will refer to this fascinating papyrus. The following puzzle has been created in the style of the puzzles in that book, focusing on fractions and ratios.

Bill, Arnie, Lisa, and Lucy went to a coffee shop last night. Lucy drank 6 pints of coffee (an awful lot, don't you think?). Bill drank half of Lucy's amount, Arnie drank half of Bill's, and Lisa drank half of Arnie's. How many pints did Lisa drink?

Incidentally, the textbook is called both the *Ahmes Papyrus*, after the Egyptian scribe Ahmes, who copied it, and the *Rhind Papyrus*, after the Scottish lawyer and antiquarian, Alexander Henry Rhind (1833–1863), who purchased it in 1858 while vacationing in Egypt. The papyrus had been found a few years earlier in the ruins of a small building in Thebes in Upper Egypt.

CLUE: PAGE 138 • ANSWER: PAGE 150

Matching Colors: 20 Balls

The following puzzle genre was invented by the late Martin Gardner (1914–2010), who was the puzzle columnist for *Scientific American* for 25 years, starting in 1956. Gardner is acclaimed as one of the modern era's best puzzle-makers.

For this kind of puzzle, you must consider the "worst-case scenario"; in other words, do not assume that you will be lucky and draw out the matching balls with your first two draws. The worst-case scenario means that you need to envision yourself unlucky to the maximum.

In a box there are 20 balls: 10 white and 10 black. The balls are equally spherical, and both weigh and feel the same. If you are wearing a blindfold, what is the least number of balls you must draw from the box to get a pair of balls matching in color (two white or two black)?

CLUE: PAGE 138 · ANSWER: PAGE 150

12

Matching Colors: 30 Balls

Let's do another one in this genre, perhaps adding some complexity to it!

In a box are 30 billiard balls—10 white, 10 black, and 10 red—scattered haphazardly. Again, they all weigh and feel the same. If you are wearing a blindfold, what is the least number of balls you must draw out to get a pair that matches: that is, two white balls, two black balls, or two red balls?

As you can see, logic can be confusing. Lewis Carroll made this rather appropriate statement, albeit rather tongue-in-cheek: "Contrariwise, if it was so, it might be; and if it were so, it would be; but as it isn't, it ain't. That's logic."

CLUE: PAGE 138 • ANSWER: PAGE 151

A Common Word

The study of how words form their own associations and semantic patterns is the basis of rhetoric, the art of using persuasive and effective language. The associative aspect of language is also the basis for various word puzzles. Consider this one below.

What single word can be added to all of the following words to create legitimate new compound words?

jail, cage, brain, call, song

The first book of crossword puzzles was published by Simon & Schuster in 1924. Titled *The Cross Word Puzzle Book,* it consisted of a compilation of crossword puzzles from the newspaper *New York World* and was an overnight best seller.

CLUE: PAGE 138 • ANSWER: PAGE 151

Fibonacci's Legacy

Leonardo Fibonacci of Pisa (1170–1250) made many original contributions to mathematics, including his legendary Fibonacci sequence. Here is a stretch of his original sequence in the form of a puzzle.

What number comes next?

1, 1, 2, 3, 5, 8, 13, 21, 34, 55 . . .

IQ tests typically use sequences similar to this one. The first widely used IQ test was prepared by French psychologists Alfred Binet and Théodore Simon in 1905. Their aim was to measure intelligence independently of social class, ethnicity, or any other social variable.

CLUE: PAGE 138 • ANSWER: PAGE 151

A Trickier Fibonacci

Sequence puzzles require a combination of math and logic. Here's another one for you to try.

What number comes next?

3, 4, 6, 9, 13, 18, 24, 31 . . .

Did you know that there is a branch of mathematics, called recreational mathematics, which studies the history, structure, and meaning of mathematical puzzles?

CLUE: PAGE 138 · ANSWER: PAGE 151

Palindrome: Vessel

A palindrome is a word or expression that reads the same backward as forward. Here's a famous example of a palindrome: *A man, a plan, a canal, Panama!* Note how adjusting the punctuation and the spaces between the words allows one to read the sentence in either direction.

What five-letter English word starting with *k* can be read forward and backward and refers to a type of vessel?

Palindromes were important in antiquity, as a famous acrostic, known as the Sator Acrostic, reveals, found among the ruins of Pompeii (and other ancient places). The word *acrostic* derives from the Greek term *akrostikhis*, which is a combination of *akron* ("head") and *stikhos* ("row, line of verse"). Literally, therefore, it meant "the line at the head," pointing out the fact that in addition to horizontal rows, an acrostic contains a vertical row formed by the letters at the "head" of each line. The word found its way into English in the 16th century, suggesting that anagrams had become popular during that period.

Palindrome: Round and Round

Now, try your hand at this palindrome.

What seven-letter English word starting with *r* can be read forward and backward, and refers to something that makes things go round and round?

Ralph Waldo Emerson (1803–1882) once wrote that "a character is like an acrostic or Alexandrian stanza; read it forward, backward, or across, it still spells the same thing." Incidentally, one of the most interesting problems in mathematics is to find palindromic prime numbers (numbers that have no factors other than themselves and 1). For example, 101 is a prime number and its digits can be read forward and backward. Another such number is 919. Can you find others?

CLUE: PAGE 139 • ANSWER: PAGE 151

Fishy Size

The following puzzle is a classic one in simple mathematics, but a tricky one nonetheless. Puzzles like this reach considerably far back in history.

The other day my friend went fishing. She caught an enormous trout that was 20 feet long plus half of its own length. What was its total length?

The term *enigmatology* was coined by Will Shortz, puzzle editor at the *New York Times,* to draw attention to the importance of studying puzzles systematically and examining them just like any other type of human intellectual or creative artifact. The word is defined simply as "the study of puzzles."

Doublet: Cold to Warm

Lewis Carroll was not only a great writer of children's books and a mathematician, but also one of the most brilliant puzzle creators in history. His "doublet" puzzle has become a staple in the category of word games. The challenge is to transform one word into another by changing only one letter at a time, forming a genuine new word (not a proper name) with each letter change. For example, turn the word *grow* into *flop* in just two steps between the beginning and ending words: GROW—(1) *glow*—(2) *flow*—FLOP. The first step is to change the *r* in *grow* to *l* to get *glow*. The second step is to change the *g* in *glow* to *f* to get *flow*. Finally, by changing the *w* in *flow* to *p*, we get the final world *flop*.

Change *cold* to *warm* in just three steps (between the starting and ending words, of course).

Lewis Carroll himself stated that he invented the doublet game on Christmas Day of 1877. Actually, the first mention of the doublet appears in an entry in his diary of March 12, 1878, in which he calls it *word links*, describing it as a two-player game. He uses the name *doublet* for the first time in the March 29, 1879, issue of *Vanity Fair*, for which he wrote many such puzzles.

CLUE: PAGE 139 · ANSWER: PAGE 151

Alcuin's Masterpiece

This puzzle is one of a set of three called the "River Crossing Puzzles," posed originally by the famous English scholar and ecclesiastic Alcuin (735–804 CE). The puzzle below—one of the greatest puzzles of all time—comes from this manual. Can you solve it?

A traveler comes to a riverbank with a wolf, a goat, and a head of cabbage. To his delight he sees a boat that he can use for crossing to the other bank, but to his dismay, he notices that it can carry no more than two—the traveler himself, of course, and just one of the two animals or the cabbage. As the traveler knows, if the two are left alone together, the goat will eat the cabbage and the wolf will eat the goat. The wolf does not eat cabbage. How does the traveler transport his animals and his cabbage to the other side intact in the minimum number of back-and-forth trips?

This puzzle is seen by mathematicians as providing the initial insight into the theory of permutations and combinations. Remember those from school?

CLUE: PAGE 139 • ANSWER: PAGE 152

Word Chains

The number of ingenious puzzles invented to challenge our knowledge of language associations is quite large, attesting to the enormous appeal that wordplay holds for us. One clever type of word puzzle sometimes found in cryptic crosswords (i.e. crosswords with trick clues) is the word chain. This format provides two words separated by a space. Your challenge is to fill that space with a word that produces a legitimate word or expression that relates to both the given words: for example, *key* _____ *link.* Inserting *chain* in the space produces two legitimate words: *keychain* and *chainlink.* That's all there is to it.

Which word inserted in the space produces two legitimate words when combined with the given words?

wild _____ *style*

The 2006 film *Wordplay*, directed by Patrick Creadon, is an enlightening documentary on the appeal of crossword puzzles focusing on Will Shortz, the *New York Times* puzzle editor and founder of the annual crossword tournament. The movie tracks Shortz's lifelong interest in puzzles and takes a look at the puzzle tournament he founded in 1978.

Count Carefully

Here's a tricky puzzle in arithmetic. I am unaware of who first devised this one, but it appears frequently in all kinds of puzzle collections and websites. Be careful!

A pencil and eraser together cost 55 cents. The pencil costs 50 cents more than the eraser. How much does the eraser cost?

Does emotion play a role in puzzles? It does. Not only do puzzles typically bring about a feeling of satisfaction from successfully finding a solution, or a feeling of frustration from not getting it—research also shows that they activate specific emotional centers in the brain.

CLUE: PAGE 139 • ANSWER: PAGE 152

Trainspotting

This is a classic puzzle that always seems to stump solvers. It has appeared in collections since at least the mid-20th century. Though it's actually an easy puzzle, it can definitely appear complicated.

A train leaves New York for Chicago traveling at the rate of 75 miles an hour. Another train leaves Chicago for New York an hour later, traveling at the rate of 50 miles per hour. When the two trains meet (cross each other), which one is nearer to New York?

CLUE: PAGE 139 • ANSWER: PAGE 152

Rebus: Popular Expression

Do you know what a rebus is? It is a form of writing that suggests a meaning or message through a combination of symbols and words. Rebus writing goes back considerably in time, but the puzzle version is a relatively modern phenomenon. It can consist of pictures, letters, and numbers in some combination, or a layout of words and symbols that hide some message. It also might involve a play on the sounds of the letters, on the location of words and symbols with respect to each other, or on the pronunciation of the symbols used. The trick is to "read" the message to literally "see" what it yields. For example, what expression does W1111HILE hide? Look closely and we can see that ones have been inserted in the word *while*—giving us "ones in while." With some minor adjustment, we arrive at the hidden phrase "once in a while." Here is a simple rebus for you to try.

What expression does the following rebus suggest?

BRIDGE
DUCK

Rebuses were popular in the Middle Ages and were used to write names, perhaps because they are easier to figure out if one is illiterate.

The Riddle of the Sphinx

No book of classic puzzles worth its salt would neglect to include the first documented riddle in history. As already discussed, according to legend, when Oedipus approached the city of Thebes, he encountered a gigantic Sphinx guarding the entrance to the city. The menacing beast confronted Oedipus, posed the following riddle to him, and warned him that if he failed to solve it, he would die instantly at the Sphinx's hands. Oedipus answered it correctly, and the rest is history.

What creature moves on four legs at dawn, two at noon, and three at twilight?

The main character of the *Saw* movie series is a figure called Jigsaw. He poses puzzles to people who must solve them in order to save their lives. In essence, Jigsaw is a modern descendant of the Sphinx.

CLUE: PAGE 139 · ANSWER: PAGE 153

Level

2

PRODIGY

Anagram: Dormitory

Throughout history, many famous personages avowed their belief in the divinatory power of anagrams. During the siege of the city of Tyre, Alexander the Great (356–323 BCE) had a dream in which a satyr appeared before him. Troubled by this occurrence, Alexander summoned his soothsayers the next morning to interpret the dream. They pointed out that the word *satyr* itself contained the answer, because the Greek word for satyr was an anagram of "Tyre is thine." Reassured, Alexander went on to conquer the city the subsequent day. The point is that anagrams may reveal truth indirectly, as the following puzzle will illustrate. It is a classic anagram, found throughout wordplay collections.

The letters of the word *dormitory* can be rearranged to produce a two-word phrase that describes a typical dormitory. What is that phrase?

To reiterate here, anagrams might be prophetic after all. An anagram of radioactivity pioneer Marie Curie (1867–1934) is "Me, radium ace." How accurate!

CLUE: PAGE 140 • ANSWER: PAGE 153

Odd-One-Out: Shapes

Recall the odd-one-out puzzle from level 1 (puzzle 8). Here's a slightly tougher one for you to try. Also bear in mind that this type of puzzle is enjoyed enormously by children learning to read and write.

Which of the words in this set does not conceptually belong?

triangle, cube, square, octagon, pentagon

Portia's Dilemma

LOGIC

Are you familiar with *The Merchant of Venice* by William Shakespeare (1564–1616)? A highlight of that play is the scene in which Portia's potential suitors have to choose among three caskets composed of gold, silver, and lead. The one who chooses the correct casket containing Portia's portrait and a scroll will win her hand in marriage. This kind of puzzle has inspired many variations. Here's one made up just for you.

Imagine three boxes: a box with two black ties in it, a second box containing two white ties, and a third box with one white and one black tie. The boxes are labeled, logically enough, *BB* (= two black ties), *WW* (= two white ties), and *BW* (= one black tie, one white tie). However, someone has switched the labels, so that now each box is labeled incorrectly. Can you determine the actual contents of each box after one draw? And from which box should you make that draw?

Here is a sage thought from Shakespeare: "There is nothing either good or bad, but thinking makes it so."

CLUE: PAGE 140 • ANSWER: PAGE 153

Weighing Conundrum

Here is a classic puzzle whose origin is unknown. Is it a math puzzle or is it a logic puzzle? Maybe both.

I have seven billiard balls, one of which weighs less than the other six. Otherwise, they all look exactly the same. How can I identify the one that weighs less on a balance scale, using that scale no more than two times?

Math and reason are certainly connected, aren't they? English mathematician James Joseph Sylvester (1814–1897) put it perfectly when he stated, "Mathematics is the music of reason."

Don't Smoke

Here's a kind of puzzle that requires some truly clever thinking. It also bears an implicit warning about the dangers of smoking.

Jack used to smoke. One day he decided to quit, cold turkey, after smoking the 27 cigarettes that remained in his pocket. He took out the 27 cigarettes and started to smoke them, one by one. Since it was his habit to smoke only ⅔ of a cigarette, Jack soon realized that he could glue three butts together to make another cigarette. So, before giving up his bad habit, how many cigarettes did he end up smoking?

On the subject of arithmetic, poet Carl Sandburg (1878–1967) offered the following witticism: "Arithmetic is where the answer is right and everything is nice and you can look out of the window and see the blue sky—or the answer is wrong and you have to start over and try again and see how it comes out this time." It rings true, doesn't it?

CLUE: PAGE 140 • ANSWER: PAGE 155

Iron to Lead

Recall Carroll's doublet from puzzle 19. Carroll later modified the rules to make his puzzles more difficult. Here is the example he used to introduce a new version of the doublet. Can you solve it? Rearranging the letters of a word is, of course, making an anagram of the word.

Change *iron* into *lead* by introducing a new letter or by rearranging the letters of the word at any step. You may not do both in the same step. Each time you must create a genuine word, of course.

According to some scholars, anagrams originated in the writings of the Greek poet Lycophron (320–280 BCE) who lived in Alexandria. Lycophron's two most famous anagrams were on the names of Ptolemy (100–170) and his queen, Arsinoë, in his famous poem on the siege of Troy titled *Cassandra*. It seems that we have been playing with words forever!

Racing: Four Runners

Recall puzzle 7. As promised, here is a slightly trickier puzzle in this genre.

Armand, Claudio, Shirley, Dina, and Elgin competed against each other in a high school race. Elgin beat Claudio, Shirley beat Elgin, Dina came in right after Shirley, and Armand beat Shirley. In which order did they all finish?

CLUE: PAGE 140 • ANSWER: PAGE 156

Lateral Thinking: Frank's Children

LOGIC

As promised, here is your second puzzle in lateral thinking. Again, you might find the same form of puzzle under a different rubric in other puzzle collections. But I certainly think it involves lateral thinking. Try it yourself and you be the judge. This one is a classic, by the way.

Frank has three daughters, each of whom has a brother. How many children did Frank have altogether?

The expression *thinking outside the box* likely comes from the way in which the famous nine-dot puzzle is solved: Connect nine dots arranged in a square alignment using four lines, without the pen leaving the paper or any dots left over. One of the first appearances of this puzzle is in a collection by puzzlist Sam Loyd (1841–1911) dating back to the late 1800s.

Missing Number

Here's another type of math puzzle that IQ test constructors love to pose. It is actually quite simple, if you know your basic math.

What number completes the following sequence?

3, 9, 27, 81, 243 . . .

CLUE: PAGE 140 • ANSWER: PAGE 156

Alphametic: Three Letters

The following puzzle is called an *alphametic*, since it replaces numbers with letters—hence a blend of *alpha(bet) + (arith)metic*. It belongs to a puzzle genre called *cryptarithmetic*, developed by two great puzzle-makers, the American Sam Loyd and the Englishman Henry Dudeney. This puzzle is actually rather simple, but if you are unfamiliar with this type, it can be quite challenging.

What numbers do the following letters stand for?

TIP + PIT = APA (A = 5)

One of Sam Loyd's best chess puzzles is the following one: Place the Queen on any chessboard square and then move her over the 64 squares and back again to the starting square in 14 moves. Can you solve it?

Alphametic: Four Letters

Now, try your hand at this one, which increases the difficulty level a notch or two.

What numbers do the following letters stand for?

SLOB + BLOL = KOOK

A Classic Nursery Rhyme

The following nursery rhyme appeared in 18th-century England. It has become a classic conundrum with a twist, so be careful!

As I was going to St. Ives, I met a man with seven wives. Each wife had seven sacks. Each sack had seven cats. Each cat had seven kits. Kits, cats, sacks, wives, how many were going to St. Ives?

Voltaire's Riddle

The following riddle was composed by the great French satirist Voltaire (1694–1778).

> **What of all things in the world is the longest, the shortest, the swiftest, the slowest, the most divisible and most extended, most regretted, most neglected, without which nothing can be done, and with which many do nothing, which destroys all that is little and ennobles all that is great?**

As seen in the Riddle of the Sphinx, riddles in the ancient world were regarded as coded messages from the gods. This is because the ancients viewed riddle talk as the gods' natural language. They therefore believed that only those equipped with special knowledge, such as the Greek oracles who spoke in riddles, could truly understand the message. In later years however, prophets such as Nostradamus (1503–1566) wrote in riddles, as well.

CLUE: PAGE 141 · ANSWER: PAGE 156

Jumble: A Proverb

This genre of puzzle was created in 1954 by Martin Naydel (1915–2006), who is better known for his comic books. It consists of the letters of words all scrambled up. The challenge is to unscramble the letters to find the original words. For example, one way to scramble the letters of the word FRIEND is DFRNEI. There are various versions of this game, including one in which you are provided with several scrambled words, and your goal is to reconstruct a phrase or expression. Here is an example for you to try.

The letters in the words of a famous proverb have been scrambled. Can you figure out what the proverb is?

WOT   WSRGON   OD   OTN   KEAM   A   GHTIR.

The English writer John Dryden (1631–1700) once wrote that anagrams "torture one poor word ten thousand ways."

40

Jumble: Alexander Pope

Here is another jumble puzzle for you.

Unscramble the letters to get a famous quotation by Alexander Pope:

OT RER SI UHAMN; OT ROFIGEV, NEVIDI.

You might find it interesting that wordplay comes in many varieties and with ingenious twists. A pangram, for instance, is a sentence using every letter of the alphabet at least once. The best-known English example is *The quick brown fox jumps over the lazy dog,* which uses all 26 letters.

CLUE: PAGE 141 • ANSWER: PAGE 157

Most Frequent Number?

This is a math puzzle found commonly in puzzle collections. Here's our version.

What digit between 1 and 1,000 (inclusive) is the most frequent?

There are many games based on mathematics. One is a two-player mathematical game called Grundy's Game, which starts with a heap of objects such as checker pieces. The two players take turns splitting the heap into two piles of different sizes. The game ends when the piles can no longer be split into different sizes, such as when one pile consists of only two pieces. Try it with a partner—you'll find it stimulating and fun. But first, look up the simple rules on the Internet.

Containers

Here is a measurement puzzle that appears in many guises in puzzledom.

Mark has a 3-gallon container, a 5-gallon container, and a 10-gallon container. He needs to measure out exactly 7 gallons of water. How does he do it?

CLUE: PAGE 141 • ANSWER: PAGE 157

The Beverley Family

As you know, this genre was invented by Henry Dudeney. There are now many variants of his original puzzle, so here's one more for you.

In the Beverley family, who is the father's father's son in relation to the Beverley's only son? Beverley is the father, by the way.

Many of Dudeney's puzzle ideas continue to surface in different versions throughout the puzzle world today. One of his best-known collections of puzzles is called the *The Canterbury Puzzles* (1907), because the collection features characters from Geoffrey Chaucer's (1343–1400) work *The Canterbury Tales* (1476).

CLUE: PAGE 141 • ANSWER: PAGE 157

Reversals

The following puzzle involves a play on words that reaches deep into puzzle history. It constitutes an interplay between spelling and meaning—much like anagrams! You are given two meanings or definitions. The first one defines the word, while the second one defines the word you will get by reversing the first and last letters. Here's an example: (1) "enjoying oneself in a noisy and lively way" and (2) "a rigid bar placed on a pivot, used to lift a heavy load." The answer is *revel/lever*. The term *revel* corresponds to meaning 1. Then switching the first and last letters of this word produces the word *lever*, which corresponds to meaning 2.

The following two meanings refer to words that have their first and last letters reversed:

(1) "guide someone" and (2) "a bargain." What are the two words?

CLUE: PAGE 141 • ANSWER: PAGE 157

The word *pun* means a play on words. As such, it is basically a type of puzzle. The word first appears in a 1662 work by John Dryden. While no one knows its true source, it could well have been coined by Dryden himself. The word *punlet* was used in 1819 by Samuel Taylor Coleridge (1772–1834) and *punkin* by Henry James (1843–1916) in 1866, both with the same meaning as *pun*. These terms may well be alterations of the Italian word *puntiglio*, "pique," which became *punctilio* in English. Whatever the case, the word has given birth to some derivatives of its own, such as *punnigram*, modeled on *epigram*, and *punnology*, modeled on *etymology*.

Counting Pets

Conundrums that involve tricky counting go all the way back to antiquity. Mathematicians have invented them throughout history to play around with numbers. Here's one for you that tests your cleverness in counting.

How many pets does my friend Mary have if all of them are dogs except two, all are cats except two, and all are rabbits except two?

CLUE: PAGE 141 · ANSWER: PAGE 157

Counting Coins

As we have seen before, puzzles with coins are frequent in the universe of puzzle-making. Here's yet another one.

Alex has 20 coins in his pocket, consisting of dimes and nickels. Altogether the coins add up to $1.35. How many of each coin type does he have?

Rebus: Book Title

Recall the rebus puzzle we saw before (puzzle 24). Here's a trickier one.

What book title does this rebus stand for?

(The) HCATAT

Incidentally, the French satirist Voltaire loved sending cards and letters composed as rebuses. Here's one that he might have appreciated, if he knew this familiar English saying.

MO**ONCE**ON

CLUE: PAGE 141 • ANSWER: PAGE 158

Phillips's Liar Puzzle

In the 1930s, the British puzzlist Hubert Phillips (1891–1964) added a provoking new category to the logic puzzle genre, which can be called liar puzzles. Phillips was known among his readers as "Caliban," the monstrous figure in Shakespeare's play *The Tempest*. Below is a puzzle derived from the type concocted by Phillips.

Five people were questioned yesterday by the police. One was suspected of having murdered a romantic rival. Here's what each one said:

EILEEN: **Earl is the murderer.**

EMMA: **Yes, Earl is the murderer.**

EUGENIA: **Earl is, without a shadow of a doubt, the murderer.**

EDWIN: **Emma lied.**

EARL: **Emma told the truth.**

Four statements were lies and one was true—strangely, one by the murderer. Can you identify the murderer?

What Did He Say?

The following puzzle is a paraphrased version of a classic nut credited to the same clever Caliban who invented liar puzzles. It never fails to stump those who come across it for the first time. Can you solve it?

The people of an island culture belong to one of two tribes: the Bawi or the Mawi. Since they look and dress alike, and speak the same language, they are virtually indistinguishable. It is known, however, that the members of the Bawi tribe always tell the truth, whereas the members of the Mawi tribe always lie. Dr. Mary Brown, a linguist, recently came across three men. "To which tribe do you belong?" she asked the first man. "Duma," he replied in his native language. "What did he say?" asked Dr. Brown of the second and third men, both of whom had learned to speak some English. "He said that he is a Bawi," said the second. "No, he said that he is a Mawi," said the third. Can you figure out the tribes to which the second and third men belonged? And is it possible to determine the first man's tribe?

CLUE: PAGE 141 · ANSWER: PAGE 158

Nabokov's Word Golf

Recalling Carroll's doublet puzzle (puzzle 19), the following teaser was created by the mad narrator in Vladimir Nabokov's novel, *Pale Fire* (1962), who calls it "word golf." Here is Nabokov's version of the puzzle. Can you solve it?

Change *lass* to *male* with three steps (or word links) between.

Before the Internet and online games, one of the most widely played games was solitaire, a card game played without a partner. The object is to form sequences of cards in either ascending or descending order (or sometimes both). There is little challenge to winning most solitaire games, mainly luck—though some games are in fact quite challenging since the way in which they are laid out guides the solution strategy. Therefore, many play solitaire not only to while away idle time, but to see if luck is on their side.

Level

3

BRAINIAC

Gardner's Matching Shoes

Remember the earlier two puzzles (11 and 12) that involved drawing balls of different colors with the goal of obtaining a match? Well, here is a trickier version of this type of puzzle.

Assume a box contains 6 pairs of black shoes and 6 pairs of white shoes, all mixed up. What is the least number of shoes you must draw—with a blindfold on—to be sure of getting a matching pair of black or white shoes?

The Snail's Journey

MATH

The following puzzle is one of the oldest puzzles around. It turned up for the first time in the third section of Leonardo Fibonacci's *Liber Abaci* (Book of the Abacus) of 1202. In that book, the animal involved was a lion. The snail version, on the other hand, can be traced to an arithmetic textbook written by Christoff Rudolff and published in Nuremberg in 1561. It is this version presented below.

A snail is at the bottom of a 30-foot well. Each day it crawls up 3 feet and slips back 2 feet. At that rate, on what day will the snail be able to reach the top of the well?

Claude Gaspard Bachet de Méziriac (1581–1638) was a great mathematician whose compilation of puzzles, *Problèmes plaisants et délectables,* was published in 1612. This volume represented the first anthology of challenging math puzzles collected in a systematic way, many of which continue to exist today in different guises and versions.

CLUE: PAGE 142 • ANSWER: PAGE 159

Anagram: Princess Diana

As you now know, it was historically believed that the anagram of a name contained the name-bearer's destiny. Mary Queen of Scots (1542–1587), who died by execution, was posthumously memorialized with the Latin expression *Trusavi regnis morte amara cada* ("Thrust by force from my kingdom I fall by a foul death"), which is an anagram of *Maria Steuarda Scotorum Regina* ("Mary Stewart Queen of Scots") (Note that V = U in Latin script). Eerie, isn't it? The puzzle below is a modern-day example of a prophetic anagram.

By rearranging the letters of "Princess Diana" you will arrive at a phrase that describes how she died. Lewis Carroll was particularly gifted in crafting this type of apt tribute. His anagrams on the name of British humanitarian Florence Nightingale (1820–1910) constitutes a fitting eulogy: *Florence Nightingale = Flit on, cheering angel!* **Another on the name of British political agitator William Ewart Gladstone (1809–1898) makes an appropriate commentary on his firebrand personality:** *William Ewart Gladstone = Wild agitator! Means well!*

Racing: Five Runners

In the style of the first two racing puzzles in this book, numbers 7 and 32, here is your third challenge of this category.

Five runners competed at a yearly racing meet. Rashad came in right after Mary and just ahead of Jack. Walter came in immediately after Jack, and Thomas beat Mary. Who came in first?

CLUE: PAGE 142 • ANSWER: PAGE 159

Anagram: Credible

Here is a twist on the anagram puzzle genre.

If you put together the words *livable* and *bee* and anagrammatize them, you will get a word meaning "credible."

Did you know that bees make their hives with hexagonal patterns because this is the most efficient geometrical figure for coverage of the area? Does this make bees puzzle-solvers? It depends on how we define puzzles, doesn't it?

CLUE: PAGE 142 • ANSWER: PAGE 159

Anagram: Space Traveler

Here is another anagram of the same type.

If you put together the words *roast* and *tuna* and anagrammatize them, you will get a word meaning "space traveler."

A well-known anagram of Clint Eastwood's name, which describes his early acting career, is *Old West action*.

CLUE: PAGE 142 · ANSWER: PAGE 160

Lateral Thinking: Truck Stuck

As promised, here is your third and last puzzle in this genre. It is *the* classic conundrum, found in virtually all treatments of lateral thinking, from technical to just-for-fun.

A big truck got stuck under a bridge that was too low to pass beneath. The driver had underestimated the truck's height. He brought in help and no one could figure it out without destroying the truck or the bridge. Then a child came along with a simple resolution to the problem. What was the child's solution?

Word Combinations

Word games exist in a variety of formats and styles, as we have seen. Here's one made up especially for you.

Take one letter from each of the following five words. When combined in order of the current list, these letters produce a new five-letter word meaning "correct":

trip, chin, enough, sigh, temper

Some crosswords allow for more than one correct answer to the same set of clues. Interestingly, they are called "Schrödinger crosswords," alluding to the quantum physicist Erwin Schrödinger's (1887–1961) famous thought experiment known as "Schrödinger's Cat," in which a cat may be simultaneously both alive and dead.

CLUE: PAGE 142 • ANSWER: PAGE 160

How Many Socks?

The origin of this mathematical teaser remains a mystery, but it shows up in classic collections all the time.

There are somewhere between 50 and 60 socks in a box. If you count them 3 at a time, you will find that 2 socks are left over. Alternatively, if you count them 5 at a time, you will find that 4 socks are left over. How many socks are in the box in total?

The great writer Lord Byron (1788–1824) once wrote the following witticism: "I know that two and two make four—and should be glad to prove it too if I could—though I must say if by any sort of process I could convert 2 and 2 into *five* it would give me much greater pleasure."

Caliban's Truth-Teller or Liar?

Hubert Phillips's marvelous truth-teller-versus-liar puzzle is a favorite of mine. I just can't get enough of this challenge myself, so here's one more for you.

In a certain village, people belong to either a truth-telling or a liar clan. Dr. Brown, our linguist, ran into three men from that village, named Tor, Dor, and Gor.

BROWN: **Tor, is Dor a truth-teller?**
TOR: **Yes.**
BROWN: **Dor, do Tor and Gor belong to the same clan?**
DOR: **No.**
BROWN: **Gor, is Dor a truth-teller?**
GOR: **Yes.**

To which clan did each of them belong?

The *Codex Cumanicus* was a medieval handbook designed to help Catholic missionaries communicate with the Cumans, a nomadic Turkic people. For some reason, the book contained many riddles.

CLUE: PAGE 142 · ANSWER: PAGE 161

Sequence: Seven Letters

LOGIC

This kind of puzzle finds its way into any classic repertoire of word games. It is similar to puzzle 9 presented earlier on the topic of lateral thinking.

What letter comes next?

OTTFFSS...

Here is a thought-provoking and very relevant fact. Because of promising research that puzzles help stave off memory loss, it is estimated that in the next few years the "brain fitness industry" will be worth over two billion dollars. So puzzles can indeed be profitable! But this does not deny their effectiveness and health value. While research has linked brain shrinkage to Alzheimer's disease, studies also show that physical exercise combined with puzzle-solving constitutes a powerful united strategy for preventing brain shrinkage.

CLUE: PAGE 142 · ANSWER: PAGE 161

Sequence: Eleven Letters

Here's another "letter code" puzzle for you to decipher.

What letter comes next?

JFMAMJJASON . . .

No strict correlation has ever been found between IQ and ability to solve puzzles. This is good news, since it means that anyone, no matter how high or low their IQ, can do puzzles and enjoy them. Puzzles are effective for enhancing brain functioning because they stimulate the imagination and logical parts of the brain in tandem. Edgar Allan Poe called it the human being's "bi-part soul."

CLUE: PAGE 142 • ANSWER: PAGE 161

Sequence: Six Letters

Now that you have gotten the hang of it, let's do one more puzzle like number 62. As they say, practice makes perfect.

What letter comes next?

APIWAT...

Legend has it that the ancient Greek poet Homer's death was precipitated by his distress at his failure to solve the following riddle posed to him by a group of fishermen: "What we caught, we threw away. What we could not catch, we kept." The answer is "fleas."

Code Logic

Codes have been around since antiquity, as we have already discussed. Here's one more for you to crack.

The following sequence of digits hides an important American event:

741776

Can you figure out which one?

CLUE: PAGE 143 • ANSWER: PAGE 162

How Much Does It Weigh?

MATH

Weighing challenges have always existed in human societies, so here's another timeless weighing puzzle for you solve.

Nora placed a brick on one pan of a weighing scale. The scale balanced when she placed ¾ of another brick of the same kind plus ¾ of a kilogram weight on the other pan. How much did the original brick weigh?

A specialist in mechanical puzzles, the American Jerry Slocum, founded the Slocum Puzzle Foundation in 1993, a nonprofit organization aiming to educate people on the importance of puzzles.

CLUE: PAGE 143 · ANSWER: PAGE 162

Change-a-Letter: Perspicuity

In this word game you must construct one word from another, simply on the basis of these two words' definitions.

Change a letter in the word for "monetary aid to help someone" to produce a new word meaning "perspicuity."

The English word *riddle* comes from an Old English word meaning "sieve," indicating that a riddle is indeed a puzzle from which one has to "sift out" its meaning.

CLUE: PAGE 143 • ANSWER: PAGE 162

A Bigger Fibonacci

Recall the Fibonacci sequence. Here is another puzzle based on the same principle.

What number logically comes next?

2, 4, 12, 48, 240, 1440 . . .

Many discoveries in mathematics surface from puzzles. A puzzle known as the Königsberg Bridge Problem, devised by the mathematician Leonhard Euler (1707–1783), influenced the development of modern-day graph theory and topology, both of which are crucial for many scientific disciplines. The puzzle basically asks you to traverse a network of seven bridges without repeating any portion of your route. This puzzle is available on numerous websites if you'd like to try it yourself.

Which Colors Make a Couple?

Recall puzzle 5, Dudeney's marvelous deduction puzzle. These are now very popular and classified under the general rubric of "logic puzzles." Here's another one for you.

> Last week, three couples made dates to go to the prom. One girl was dressed in red, one in green, and one in blue. The boys also wore outfits in the same three colors. While the three couples were dancing, the boy in red said to the girl in green and her partner: "Not one of us is dancing with a partner dressed in the same color." Can you tell who went with whom to the prom?

Here's an interesting fact: One of the best-selling puzzle-games of all time is the Rubik's Cube, which has sold over 500 million copies. Have you ever played with the Cube? Solving it requires a combination of logic and visual reasoning in large doses, so I always find it difficult to crack, but some people can achieve it literally in seconds.

CLUE: PAGE 143 · ANSWER: PAGE 162

Sequence: Four Words

By this point in the collection, you have solved a few number sequence puzzles. Now that you've familiarized yourself with that format, it's time to try your hand at word sequences.

Which word logically comes next?

spot, tops, pots, opts . . .

70

Lies, Lies

Recall the various liar puzzles you have solved so far, including 48, 49, and 60. Most of them are descendants of the original prototype by Hubert Phillips. Here's another one to keep you on your toes.

Yesterday a bank was robbed. Four suspects were rounded up and interrogated. One of them was indeed the robber. Here's what they all said under questioning by the police:

ALEX: **Daniela did it.**
DANIELA: **Tara did it.**
GARY: **I didn't do it.**
TARA: **Daniela is a liar. I am innocent.**

Only one of these four statements turned out to be true; the others were false. Can you figure out who the robber was?

CLUE: PAGE 143 · ANSWER: PAGE 163

Signs, Signs

This type of puzzle was a favorite of some of the great puzzlists, including Loyd and Dudeney. It can certainly be a head scratcher.

Provide the missing signs (+, −, ×, ÷) that would make the numbers provided work in the following equation:

34 ? 43 ? 6 = 71

Here's another interesting fact about mathematicians and their love of games. Invented by Danish mathematician Piet Hein (1905–1996), Hex is a board game played on a hexagonal grid. Players take turns placing a stone of their color on a single cell of the board. The objective is to form a connected path of stones before one's opponent does. Apparently, the game was also invented independently by John Forbes Nash Jr. (1928–2015), the subject of the movie *A Beautiful Mind* (2001).

Math Signs

Here's a similar puzzle requiring the insertion of math signs.

Provide the missing signs (+, −, ×, ÷) that will lead to a correct equation:

72 ? 8 = 7 ? 2

Sometimes mathematical puzzles show up in the most unexpected places. The Monty Hall Problem was named after Monty Hall, the host of the TV quiz show *Let's Make a Deal*. It goes like this: Suppose you are given the choice of three doors. Behind one door is a car; behind each of the other two doors is a goat. You pick, say, door number 1, but the host, who knows what's behind the doors, opens a different door, number 3, which is hiding a goat. He then asks you, "Would you like to pick door number 2 instead?" What would you say? Is it to your advantage to switch your choice? In fact, it is. It turns out that those who change their answer have a two-thirds chance of winning the car, while those who stick to their choice have only a one-third chance.

CLUE: PAGE 143 · ANSWER: PAGE 163

Change-a-Letter: Paramour

This type of puzzle is a favorite of many puzzle-makers. Here's one more of this kind for you to try.

Change one letter in a word referring to a glandular organ and you will get a word meaning "paramour."

Puzzles that trick us can also frustrate us. Three posts, colored red, white, and blue cast a shadow. Which shadow is the darkest? Shadows are, in fact, all the same hue (dark). But the answer is frustrating if you didn't figure it out yourself, right?

Change-a-Letter: Caring

Here's another change-a-letter puzzle.

Change one letter in a word referring to an emitter and you will arrive at a word meaning "caring."

CLUE: PAGE 143 • ANSWER: PAGE 163

Musicians

Recall the brilliant inventions of Henry Dudeney. As we've already seen in several cases (such as puzzles 4, 5, and 68), he intended these puzzles to test our logic skills in ingenious ways. Here's another one in a similar vein, with the difficulty level raised a notch or two.

Bernard, Peter, Rhonda, and Selena are musicians. One is a drummer, one a pianist, one a singer, and one a violinist, though not necessarily in that order. Selena and Bernard play often with the violinist. Selena also plays and performs often with the drummer. Both Peter and the violinist have attended many concerts of the pianist. Peter is not the drummer. What is each person's musical field?

Puzzles of all kinds are found in literature, the arts, and even music. An example of the latter is Edward Elgar's (1857–1934) beautiful *Enigma Variations,* a work that has been characterized as a "musical cryptogram." Its unknown solution still challenges us today, as many have come forward with plausible explanations.

Level

4

MASTER-MIND

Zeno's Conundrum

The following puzzle goes back to ancient times. Historians attribute it to Zeno of Elea, who lived in the fifth century BCE and was the originator of the well-known paradoxes of motion, four of which have survived and have become important in the development of both logic and mathematics.

Consider two coins of equal size, A and B, touching each other. If B remains fixed and A is rolled around B without slipping, how many revolutions will A have made around its own center when it returns its original position?

Zeno's class paradox states that a runner who begins running at a starting line will never reach the finish line, because the runner must cover half the distance first; then from that half distance point, the runner must traverse another half distance, and from that new position another half distance, and so on ad infinitum. Conclusion? The half-distance intervals become smaller and smaller, but they go on forever. Using this reasoning, which is perfectly logical, the runner will never cross the finish line, even though, in reality, we know that the runner will indeed cross it (unless some other factor happens to stop the runner along the way).

CLUE: PAGE 143 • ANSWER: PAGE 164

Ladder Rungs

This puzzle has become a classic one, even though no one knows who invented it as far as I can tell. It reminds us in a fundamental way of the snail problem (puzzle 52), doesn't it?

During a warehouse fire, a firefighter stood on the middle rung of a ladder, pumping water into the burning warehouse. A minute later, she stepped up 3 rungs, and continued directing water at the building from her new position. A few minutes after that, she stepped down 5 rungs, and from this new position continued to pump water into the building. Half an hour later, she climbed up 7 rungs and pumped water from that new position until the fire was extinguished. She then climbed the remaining 7 rungs up to the roof of the warehouse. How many rungs were on the ladder?

Greek philosopher Pythagoras (570–495 BCE) offers us the following cogent thought, which is relevant to everything in this book: "Reason is immortal, all else mortal."

CLUE: PAGE 144 · ANSWER: PAGE 164

Racing: Six Runners

As promised in level 1's first racing puzzle, here is a more difficult challenge in this genre than the previous three (puzzles 7,
32, and 54).

**Six runners squared off at the annual meet held in their
small town. Jerry beat Bob, who beat Paula, who beat
Sarah, who beat Tim. Lorraine came in right after Tim.
Who won the race?**

As you now know, Lewis Carroll was one of the greatest puzzle-
makers of all time, if not *the* greatest. In his book *The Game of Logic*
(1886), he introduced a type of puzzle based on logical deduction by
which a solver was expected to reach a conclusion such as "Some
greyhounds are not fat," from premises such as "No fat creatures
run well," and "Some greyhounds run well."

Number Riddle: A Warm-Up

Here is an unusual type of puzzle. It is a number puzzle in the form of a riddle, thus making it twice as hard.

I am a number between 20 and 50. If you multiply me by 3 and divide the result by 9, you'll get 11. What number am I?

There seems to be a general unstated principle of puzzles: The clearer the puzzle's guidelines, the more people like to attempt it. Thus, puzzles like crosswords and sudoku likely experience such popularity because the rules for solving them are so straightforward and easy to understand. Of course, this does not make them any simpler to solve.

CLUE: PAGE 144 · ANSWER: PAGE 165

Number Riddle: Prime

Now that you have gotten the hang of it, here's one more math puzzle in this genre. By the way, if you have forgotten this definition, a prime number is a number that is not divisible by any number other than itself and 1.

I am a prime number less than 100. If you add 2 to me, you will get the next prime number greater than me. If you multiply me by 3, you will get a number which, if you subtract me from it, will give you 34. What number am I?

In the ninth volume of his book the *Elements,* Euclid (c. 300 BCE) proved that the number of primes is infinite, even though they become scarce as we move up the number line: For example, 25 percent of numbers between 1 and 100 are prime, 17 percent of numbers between 1 and 1,000 are prime, and only 7 percent of numbers between 1 and 1,000,000 are prime.

Matching Colors: 30 More Balls

Recall Martin Gardner's famous puzzle, drawing balls from a box (puzzle 11). Here's a trickier version.

> In a box are 30 balls: 10 white, 10 black, 5 green, 3 blue, and 2 yellow. What is the least number you must draw, with a blindfold on, to get a pair of balls matching in color (2 white, 2 black, 2 green, 2 blue, or 2 yellow)?

Ernő Rubik, the creator of the Rubik's Cube, once wrote that with a good puzzle "nobody is lying." In fact, puzzles do not lie, but they may sometimes deceive us with their wording and their twists.

CLUE: PAGE 144 · ANSWER: PAGE 165

Matching Colors: 25 Balls

Now try your hand at this more complex version of Gardner's puzzle.

A box contains 25 balls: 5 white, 5 black, 5 green, 5 blue, 2 yellow, 2 brown, and only 1 red. With a blindfold on, what is the least number of balls you must draw to get a pair that match in color (2 white, 2 black, 2 green, 2 blue, 2 yellow, or 2 brown)?

Above all else, Gardner is known for his ingenious mathematical games. But he was also a renowned debunker and skeptic, writing against the pseudo-science that is frequently presented as fact rather than fiction. His 1952 book, *Fads and Fallacies in the Name of Science*, led to the establishment of the skeptical movement, which promotes scientific inquiry and the use of reasoning in examining all claims.

Number Riddle: Three Digits

Now that you have become familiar with this type of numeric puzzle, we will move on to the first of three in a new format. Here the object is to identify a mysterious number. These will really give you a brain workout, and maybe even a headache.

I am a three-digit number, whose digits follow one another in numerical sequence. If you multiply me by 2, and then subtract 1 from the result, you get 245. What number am I?

Previously, we mentioned that Dudeney invented alphametics. However, an example of a verbal arithmetic puzzle like the alphametic was recently discovered from 1864. It was found on page 349 of *American Agriculturist* magazine, volume 23, published in December of that year. But, in my view, it is unlikely that Dudeney would have had access to the magazine. Moreover, his puzzle is somewhat different. Maybe some puzzles are "archetypes" that recur throughout time and cultures. Could this be such an archetype?

CLUE: PAGE 144 · ANSWER: PAGE 166

Number Riddle: A Second Prime

Here's the next in this series of numeric riddles. This one focuses on primes.

I am a prime number less than 100. If you add my two digits and then multiply me by the number that results from it, you will get 52. What number am I?

Here are a few other fascinating facts about primes. Two primes that differ by 2, such as 5 and 7 or 17 and 19 are called twin primes; however, it is not yet known whether the number of twin primes is infinite. What can be shown is that each even number (after 2) is the sum of two primes; however, this conjecture has never actually been proven. Though many conjectures about primes continue to be unsolved, primes remain among the most fascinating numbers—they are the atomic particles of all number systems.

Number Riddle: Single Digit

Finally, here is the third and last puzzle in this mind-twisting genre.

I am a number between 1 and 5. First, add me to the next number above me. Next, multiply the result by me and add me again. Finally, add the digits in the result, and you will get a number that is twice me. What number am I?

The discovery of mathematical patterns raises some fundamental questions. Are we constituted by our nature to discover it? Do the patterns discovered by mathematicians mirror the patterns hidden in nature? One thing is certain—the discovery of a pattern is a fascinating event in and of itself, as you may have experienced throughout this book.

CLUE: PAGE 144 · ANSWER: PAGE 166

Relations: Sister's Nephew

Here is another version of Dudeney's marvelous kinship relations puzzle. It's the first of three in a row.

A married woman who has only one child looks at a photo of a man and thinks, "This man's son is my sister's nephew." What relation does the woman have to the man in the photo?

In 1893 Dudeney struck up a friendly correspondence with renowned American puzzle-writer Sam Loyd, sharing many ideas. However, a rift soon developed after Dudeney discovered that Loyd was publishing many of Dudeney's puzzles under his own name. One of Dudeney's daughters recalled that her father once equated Loyd with the devil.

Relations: Mother's Grandson

Here's the second in this clever genre.

Mary has one sister and no brothers. She herself has no children. So, what relation does Mary have to her mother's grandson?

Relations: Mother's Niece

Here's the third and final puzzle in the Dudeney relations series.

A girl looks at the photo of a woman and thinks: "This woman's daughter is my mother's niece." What relation does the girl have to the woman in the photo?

Lewis Carroll once wrote, "Who in the world am I? Ah, that's the great puzzle."

Doublet: Four to Five

Try your hand at the following doublet devised by the inventor of the genre, Lewis Carroll.

Change *four* to *five* with a minimal number of steps between the two words.

Did you know that President Bill Clinton is a crossword puzzle aficionado? He even composed an online crossword puzzle for the *New York Times* in 2007.

CLUE: PAGE 145 · ANSWER: PAGE 167

Carroll's Bag of Marbles

MATH

Lewis Carroll was truly a brilliant puzzle-maker in all genres, not just a maker of doublets. Here's another example from his cunning imagination.

> **A bag contains one marble, known to be either white or black. A white marble is dropped inside, the bag shaken, and one drawn out, which also proves to be white. What is now the chance of drawing another white marble?**

Did you know that Carroll was also a photographer at around the time that photography was invented? He took many photographs that still exist today and are appreciated as art photography. He photographed famous personages, including Lord Salisbury and Alfred, Lord Tennyson.

Doublet: Wheat to Bread

Here's yet another of Carroll's inspired doublets.

Change *wheat* to *bread* with a minimal number of steps between the two words.

Riddles are found throughout our literary tradition, attesting to their perceived importance in human life. Works of literature that contain riddles range from *Ulysses* (1922) by James Joyce (1882–1941) and *Emma* (1815) by Jane Austen (1775–1817) to *Wizard and Glass: The Dark Tower IV* (1997) by Stephen King.

CLUE: PAGE 145 • ANSWER: PAGE 167

Codes and Ciphers

LOGIC

Given the extraordinary ingenuity that has gone into inventing and breaking secret military and other codes over the centuries, it is little wonder that cryptography has given rise to a genre of puzzles called, appropriately enough, *cryptograms*. These became very popular in the 19th century after Edgar Allan Poe used a cryptogram in his story "The Gold-Bug" (1843). The most common type of cryptogram puzzle is the letter-to-letter substitution, known as a Caesar cipher, because it was Julius Caesar who apparently used this as a technique. Here's how this type of puzzle works. What three-word simple phrase does the following encode: H KNUD XNT? The hidden phrase, technically called the plaintext, is I LOVE YOU. It was encoded by replacing each letter with the letter immediately before it in the normal alphabet. So *I* was replaced with *H* (the letter just before it), *L* with *K*, *O* with *N*, and so on. That is the *code*.

continued ⟹⟶

Try your hand at the following cryptogram.

This Caesar cipher hides a quotation by Mark Twain from his short novel *The Refuge of the Derelicts* (1905):

RFCPC UYQ LCTCP WCR YL SLGLRCPCQRGLE JGDC.

Can you decipher it?

Cryptography originated as a form of secret writing, not mental recreation. The writers of sacred Jewish texts, for example, often concealed their messages by substituting one letter of the Hebrew alphabet with another—the last letter in place of the first, the second last for the second, and so on. That method of cryptography was called atbash.

Sarah's Climbing Escapade

Rate and distance puzzles populate the universe of mathematics. The first such puzzles go back to antiquity and can be found in collections such as the *Ahmes Papyrus* and Fibonacci's *Liber Abaci,* mentioned several times in this book. Here's a typical example.

My friend Sarah is a mountain climber. She can hike uphill at an average rate of 2 miles per hour, and downhill at an average rate of 6 miles per hour. If she hikes uphill and then continues downhill, without spending time at the top, what will be her average speed for the whole trip?

continued ⇒⟶

As you may know, Pythagoras loved mystery and puzzles and formed a secret society to study both, now called the Pythagorean Brotherhood—though in all likelihood this name is a mistranslation, because Pythagoras encouraged full participation by women. Late in life, he married one of his students, Theano. An accomplished cosmologist and healer, Theano headed the Pythagorean society after her husband's death, and even though she faced persecution, she continued to spread the Pythagorean philosophy throughout Egypt and Greece alongside her daughters. A basic tenet of the Pythagoreans was that each natural number held symbolic significance. They claimed, for instance, that the number 1 stood for unity, reason, and creation. This is why they believed the single horn of the unicorn possessed magical powers. Today it continues to have this meaning in many cultures, where, in the form of a cordial, it is purported to cure diseases, as well as to neutralize the poisons of snakes and rabid dogs.

Anagram: Shred

Here is an anagram with an extra twist: You are not provided with the words that need transforming from one to the other. Instead, you must start with the definitions to decipher what these words are.

Rearrange the letters in a word meaning "shred" to produce a word meaning "prominent or important."

Sir Arthur Conan Doyle (1859–1930) was intrigued by puzzles, incorporating them in several of his Sherlock Holmes mysteries. He seems to have been particularly captivated by cryptograms, following in the footsteps of Edgar Allan Poe.

CLUE: PAGE 145 · ANSWER: PAGE 168

Anagram: Incentive

Here is another anagram in the same format as the previous one. You are not given the words in question, so you have to figure them out using the given definitions.

Rearrange the letters in a word meaning "inventive" to produce a word meaning "responsive."

The New York Times, which is emblematic as the publisher of the crossword puzzle, wrote in 1924 that crossword puzzles were "a sinful waste" in futility. How the times [pun intended] have changed.

CLUE: PAGE 145 · ANSWER: PAGE 168

Anagram: Fatherly

Here is one more puzzle demonstrating this format of anagram.

Rearrange the letters in a word meaning "fatherly" to produce an adjective referring to a mother and a father.

In the 1999 movie *The Matrix,* the place between the real world and the Matrix is cleverly named Mobil. This is actually an anagram of *limbo,* a region between heaven and hell.

Gardner's Sock Puzzle

Let's do another puzzle in Martin Gardner's genre of drawing items from a box.

In a box are 30 pairs of socks: 10 green, 10 red, and 10 yellow. In each case 5 are of one orientation (for the left foot) and 5 are of the other (for the right foot). Assuming you are blindfolded, what is the least number you must draw to get a pair of socks that match in color (2 green, 2 red, 2 yellow) and orientation (a pair of green left and right socks, and so on)?

Puzzles and games of all kinds are of interest to computer scientists as a means to develop algorithms. For example, Mastermind is a popular game for two players. It is played with pegs and resembles a pen and paper puzzle called "Bulls and Cows" that potentially dates back more than a hundred years. Mastermind is particularly appealing to computer scientists, since they can create algorithms to play it.

CLUE: PAGE 145 · ANSWER: PAGE 169

An Ace or a King

In 1940, Edward Kasner (1878–1955) and James Newman (1907–1966) published their famous work, *Mathematics and the Imagination.* In the book, the two mathematicians made the following apt statement: "The theory of equations, of probability, the infinitesimal calculus, the theory of point sets, of topology, all have grown out of problems first expressed in puzzle form." The following puzzle comes from that book.

Since there are 4 aces in a deck, the probability of drawing an ace from 52 cards is $\frac{4}{52} = \frac{1}{13}$. But what is the probability of drawing either an ace or a king from the deck in one draw?

Dudeney's Kinship Masterpiece

Recall from puzzle 4 that the inventor of complicated kinship puzzles was Henry Dudeney. His original invention is now considered a masterpiece in the genre. Here it is (slightly modified)—see if you can crack it.

A boy is looking at a photo: "Brothers and sisters have I none, but this man's son is my father's son." Who is the person in the photo?

Well-known writer Marilyn vos Savant, also known for having the highest recorded IQ, gives the following good advice about crossword puzzles: "People who work crossword puzzles know that if they stop making progress, they should put the puzzle down for a while." This sound advice applies to puzzles of all kinds.

CLUE: PAGE 145 · ANSWER: PAGE 169

Time Logic

This next puzzle will truly test your wits.

> **Jason works every third day at a department store as a part-time sales clerk. Alicia also works there, but only on Saturdays. The store stays open seven days a week. This week, Jason worked on Monday, October 1. On what date will the two be working together?**

Before embarking on the toughest puzzles in level 5, you might wonder: Is the ability to solve puzzles specific to humans? Maybe not. Incredibly, researchers found in 2010 that an ant species successfully solved the three-disk version of the Tower of Hanoi puzzle, invented by French mathematician Édouard Anatole Lucas (1842–1891). The puzzle uses three rods and a stack of disks in increasing sizes, which can slide onto any of the rods. It begins with the disks in increasing order, from top to bottom, on the

continued ⟶

left-most rod—the smallest disk on top and the largest on the bottom. The goal is to move all the disks to the third rod on the right, according to three rules:

1. Only one disk can be moved at one time.

2. Each move consists of taking the upper disk from one of the stacks and placing it on another rod or stack.

3. No disk can be placed on top of a smaller disk.

With three disks, the puzzle can be solved in seven moves. It is truly amazing that ants were able to solve this!

Level

5

GENIUS

Ancient Math

The *Ahmes Papyrus,* as discussed earlier, is one of the first known collections of math problems and puzzles. Since what we have is a version copied by a scribe, we know it was written even earlier than 1650 BCE. Among its puzzles, one finds conundrums dealing with fractions and early equations. The following puzzle has been created in the same spirit of the papyrus. The idea is to connect the given numbers with the relevant arithmetical signs into an equation. You must use all the numbers. Before you try solving it, here's an example. The numbers 13, 75, 248, and 4 can be combined to form a legitimate equation as follows: 4(75 - 13) = 248.

How can the numbers 10, 0, 22, and 12 be logically combined into an equation, using each number only once?

CLUE: PAGE 146 • ANSWER: PAGE 170

A Polybius Cipher

A popular type of cryptogram puzzle is the number-to-letter cipher, known as a Polybius cipher because its invention is attributed to the Greek historian Polybius (200–118 BCE). One simple Polybius code would be to replace each letter of the plaintext with digits in numerical order. For example, if the plaintext is *I LIKE LIFE*, the code would replace *I* with *1* being the first letter in the text, *L* with *2* being the second letter in the text, and so on. The end result is the ciphertext 121342154. Note that the same number is used for the same letter, no matter where it appears.

The following cipher encodes a quotation by American author Erica Jong from her book *Fear of Flying* (1973):

20 12 8 8 18 11   18 8   7 19 22   12 11 18 26 7 22

12 21   7 19 22   12 11 11 9 22 8 8 22 23.

Can you decode it?

The first evidence of the use of cryptograms for recreational purposes dates to a ninth-century manuscript found in Bamberg, Germany, which contains a cryptogram puzzle that transposes Latin letters into Greek.

Trainspotting, Logically Speaking

The following is a tricky puzzle in computation. I am not sure where the prototype for the puzzle comes from, but it appears in different guises in many classic collections, so I've included it here.

A train leaves New York for Washington every hour on the hour. Similarly, a train leaves Washington for New York, but it does so every hour on the half-hour. The trip takes five hours each way. If you are on the train from New York bound for Washington, how many of the trains coming from Washington going toward New York would you pass?

The great composer and crossword puzzle-maker Stephen Sondheim once said, "The nice thing about doing a crossword puzzle is, you know there is a solution." Maybe the fact that puzzles have solutions—unlike many other things in life—is what makes them so addictive.

CLUE: PAGE 146 • ANSWER: PAGE 170

Which Offer?

The following is a classic nut in computation—I'm unaware of this one's origin, as well, but it is found all over the puzzle universe. Here's your chance to solve it.

> You are offered a part-time job as a pizza delivery person, working only on weekends. Your boss gives you a choice of the following two salary options: (1) $4,000 for your first year of work, and a raise of $800 for each year after the first, or (2) $2,000 for your first six months of work, and a raise of $200 every six months thereafter. Which is the better offer?

Listening to classical music, such as to the symphonies of Wolfgang Amadeus Mozart (1756–1791), seems to reap many cognitive benefits. This is known, rather appropriately, as the "Mozart Effect." Puzzles can also enhance our intellectual skills and can thus be viewed, analogously, as producing the "Puzzle Effect."

CLUE: PAGE 146 • ANSWER: PAGE 171

Duma or Ruma?

Recall Hubert Phillips's clever logic puzzle 49. This one is somewhat more difficult, but the reasoning is the same.

Recall that in the village described in puzzle 49, the members of the Bawi clan always tell the truth while those of the Mawi clan always lie. While visiting this village, Dr. Brown ran into a woman and a man. They spoke a different dialect of the village language than the one spoken by her previous informants. She asked the woman, "Are you a truth-teller?" "Ruma," she replied in her dialect. Dr, Brown then asked her partner, a man who spoke English, what she had said. "She said yes," he replied, "but she is a liar." Can you figure out the clan(s) to which this man and woman belonged?

Most historians of philosophy maintain that one of the founders of logic as a reasoning system was Aristotle (384–322 BCE), even though similar concepts have been found before Aristotle and in other parts of the world. He was certainly the inventor of the syllogism, which displays how some aspects of logical thinking unfold. For example: (A) All cats are mammals; (B) Pumpkin is a cat; (C) Conclusion: Pumpkin is a mammal.

CLUE: PAGE 146 · ANSWER: PAGE 172

What Day Is It?

Remember the liar-detection puzzles from before? Well, here is one real tough nut for you in this genre. However, it may not be so tough by now, since you might have gotten the hang of it.

Alma, Barb, Charlene, Dina, Emma, and Fanny are good friends, but they love to argue—literally over the time of the day! Here they are arguing over what day of the week it is, only confusing the issue further.

ALMA: Yesterday was Friday.

BARB: Today is Saturday.

CHARLENE: No, today is not Saturday. Nor is it Monday or Wednesday.

DINA: The day after tomorrow is Tuesday.

EMMA: Tomorrow is Wednesday.

FANNY: Tomorrow is Friday.

Only one of their statements is true. All the others are false. Can you determine which day of the week it is?

continued ➤⟶

One of the most fascinating puzzle books ever written has the enigmatic title *What Is the Name of This Book?* (1978) by the late logician and mathematician Raymond Smullyan (1919–2017). It is a fascinating book because it allows readers to grasp a famous theorem in logic (called Gödel's theorem), without any background in this complicated field, simply by solving the logic puzzles that Smullyan provides.

107

Racing: Five International Runners

As promised, here is your final puzzle in this genre (see also 7, 32, 54, 78).

Five runners competed at an international race. On their shirts they bore the numbers 1, 2, 3, 4, and 5. No runner finished according to his or her number; that is, the one who wore number 1 did not come in first, the one who wore number 2 did not come in second, and so on. Number 1 and number 5 came in one after the other. Numbers 3, 4, and 5 finished one after the other, with 3 coming in just ahead of 4, who came in after 5. Number 3 was not the winner. So, who won the race?

The Sudoku Cube, created by Jay Horowitz in 2006, is a version of the Rubik's Cube. In this case, the faces or facets have the numbers one to nine on the sides instead of colors. The object is to solve the puzzle as you would sudoku, while keeping track of the cube structure, as well.

Throwing a 6 or 7

Recall puzzle 98, taken from *Mathematics and the Imagination* by Edward Kasner and James Newman. This puzzle also comes from that book.

What is the probability of obtaining either a 6 or a 7 when throwing a pair of dice?

Augustus De Morgan (1806–1871) was a famous mathematician and founder of modern set theory. For puzzle aficionados, however, he is known as author of the marvelous compilation of challenging puzzles, *A Budget of Paradoxes* (1872), which distinguishes him as one of history's great puzzle-makers.

CLUE: PAGE 146 · ANSWER: PAGE 174

109

Horace Walpole's Riddle

The following riddle comes from the English politician, writer, and man of letters, Horace Walpole (1717–1797). He is probably best known as the author of *The Castle of Otranto* (1764), identified by most literary historians as the first Gothic novel. This classic is a tough nut to crack!

Before my birth I had a name,

But soon as born I chang'd the same;

and when I'm laid within the tomb,

I shall my father's name assume.

I change my name three days together,

yet live but one in any weather.

Who am I?

Riddles have been perceived as tests of intelligence since antiquity. Known as riddle games, these appear in many literary and musical forms. For example, there is a riddle scene in Richard Wagner's (1813–1883) *Siegfried* (1857) and a riddle game in J. R. R. Tolkien's (1892–1973) *The Hobbit* (1937).

CLUE: PAGE 146 · ANSWER: PAGE 174

110

Alcuin Revisited

Recall puzzle 20, which was created by Alcuin in the 700s. River-crossing puzzles have turned out to be much more than mere exercises in logical thinking. Mathematical historians trace the conceptual roots of mathematical combination patterns to Alcuin's "River Crossing Puzzles." It is easy to recognize the roots of modern-day systems analysis, which is based on critical decision-making logic, in these simple, yet intriguing paradigmatic puzzles. Below is a version of the same concept, devised just for you.

The traveler from puzzle 20 reaches the same riverbank, with the same boat there. Along with him are his wolf, goat, head of cabbage, and this time, a mythical monster called the Wolf-Eater. The Wolf-Eater eats only wolves. In addition, when the Wolf-Eater is present on either side, he intimidates the goat, who will then not eat the cabbage. How does the traveler get them all across the river safely?

The first English translation of Alcuin's *Propositiones ad Acuendos Juvenes* was produced by John Hadley and David Singmaster and published in volume 76 of *The Mathematical Gazette* in March 1992.

CLUE: PAGE 146 · ANSWER: PAGE 174

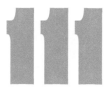

Doublet: Flour to Bread

Some of Carroll's doublet puzzles are undeniably challenging, as you will see in the following three puzzles that come directly from his pen!

Change *flour* to *bread* with the minimum number of steps between these two words.

Carroll was truly a genius. Incredibly, the fact that Carroll suffered from chronic migraines, epilepsy, stammering, and partial deafness did not deter the quickness of his intellect. It is recorded that he could write 20 words per minute, a page of 150 words in seven and a half minutes, and 12 pages in two and a half hours. He was a prolific letter writer, often writing more than 2,000 letters in a single year. He sometimes wrote backward, as well, forcing his reader to hold the letter up to a mirror in order to comprehend it.

112

Doublet: Black to White

Here's a second classic example in this sophisticated trio.

Change *black* to *white* with the minimum number of steps between these two words.

Despite being a brilliant mathematician and an unparalleled maker of ingenious puzzles, Carroll could not balance his bank account or carry out other everyday chores that we all take for granted. He often let his account go into overdraft, though he would pay it back faithfully on payday.

CLUE: PAGE 147 • ANSWER: PAGE 175

Doublet: River to Shore

And now, the third doublet to complete the series.

Change *river* to *shore* with the minimum number of steps between these two words.

114

Roman Numerals

This puzzle is a clever invention, although I am not sure of its original creator.

There is one four-digit number, and only one, which, when converted to a Roman numeral, becomes a common three-letter English word meaning "blend." What number is that?

Did you know that the Rubik's Cube appears throughout popular culture? It makes an appearance, for example, in the Spice Girls' "Viva Forever" music video.

CLUE: PAGE 147 · ANSWER: PAGE 176

House Numbers

This type of challenging puzzle particularly appeals to math educators. It requires some exact thinking.

There are 50 houses in a row, numbered consecutively from 1 to 50. How many times must the digit 1 be used?

Missing Letters

This puzzle was invented by math wizard Theoni Pappas, as far as I can tell. It may look simple, but it is truly a brain teaser.

What two letters are missing in this set:

A, H, I, M, O, T, U, V, W, ?, ?

Following the invention of mass printing technology in the 1400s, collections of riddles were among the first books printed for popular entertainment.

117

Doublet: Winter to Summer

Here is one final doublet from the pen of Lewis Carroll. It's a monster.

Change *winter* to *summer* with the minimum number of steps between these two words.

Computer scientist, mathematician, and puzzlist Donald Knuth carried out a computer study of doublets, since he believed that three-letter doublets were too easy, even though Carroll found it required six steps to turn *ape* into *man*. He also felt that six-letter doublets were trivial, since relatively few six-letter words could be linked in the step-wise fashion stipulated by the doublet. Therefore, he identified the five-letter doublet as ideal. He used a collection of 5,757 common English five-letter words, determining when two words in the set could be linked in a word ladder via other words. Knuth discovered that most of the words in that set were semantically related to each other—an amazing discovery if you think about it! He also discovered that 671 of the words could not be used to form a word ladder with any others. He called these words "aloof"—one of the actual words in that set—which means, of course, "detached."

CLUE: PAGE 147 • ANSWER: PAGE 176

118

Relations: Daughter's Mother

We've noted on several occasions throughout this collection Henry E. Dudeney's distinction as one of the great puzzle-makers of all time. Before we close, here's one last puzzle in his honor.

A woman is an only child and has herself only one child. She is looking at the photo of a woman: "That woman is my daughter's mother." Who is the woman in the photo?

In the introduction to his book, *The Canterbury Puzzles*, Dudeney wrote the following about puzzles: "Theologian, scientist, and artisan are perpetually engaged in attempting to solve puzzles, while every game, sport, and pastime is built up of problems of greater or less difficulty. The spontaneous question asked by the child of his parent, by one cyclist of another while taking a brief rest on a stile, by a cricketer during the luncheon hour, or by a yachtsman lazily scanning the horizon, is frequently a problem of considerable difficulty. In short, we are all propounding puzzles to one another every day of our lives—without always knowing it."

CLUE: PAGE 147 • ANSWER: PAGE 176

119

Number Relations

Below is one final puzzle in this category of number game.

The following four numbers can be connected in only one way into an equation:

0, 13, 17, 30.

Can you figure it out?

Here's something else that Dudeney wrote in his introduction to *The Canterbury Puzzles* that is of relevance: "Puzzles can be made out of almost anything, in the hands of the ingenious person with an idea. Coins, matches, cards, counters, bits of wire or string, all come in useful. An immense number of puzzles have been made out of the letters of the alphabet, and from those nine little digits and cipher, 1, 2, 3, 4, 5, 6, 7, 8, 9, and 0. It should always be remembered that a very simple person may propound a problem that can only be solved by clever heads—if at all."

Anagram: Opposition

Remember the tricky type of anagram puzzle based on definitions I recently called the "final" one? Well, I take that title back. I've thought it over and am adding two more. These are really challenging.

Rearrange the letters in a word meaning "opposition to" to get a word meaning "family trees."

Scientists such as Galileo, Christiaan Huygens, and Robert Hooke sometimes recorded their findings and empirical discoveries in anagram form to prevent others from stealing the credit.

CLUE: PAGE 147 • ANSWER: PAGE 176

Anagram: Marked Down

This one really is the final anagram—I promise. Can you solve it?

Rearrange the letters in a word meaning "inferences" to get a word meaning "marked down."

A blanagram is a word that is the anagram of another word except for one letter. An example is the following. The word *claw* has no anagram, but changing the *c* to *k* produces *walk*.

Number Formation

This type of math puzzle is a favorite of math teachers the world over. It is yet another classic nut whose origin is unknown.

With six 1s and three plus signs, create a formula that will add up to 24.

Math teachers are also fond of magic squares. These are arrangements of numbers that add up to the same constant sum in the rows, columns, and diagonals of a square. Try to put the first nine digits into a three-by-three square (consisting of nine cells) so that they are placed according to this rule. You'll likely find this exercise both challenging and fun. Magic squares have captured the fancy of many artists, architects, and writers. For example, a magic square is sculpted into the façade of the Passion in the Sagrada Familia church in Barcelona. The sculptor is the Catalan artist Josep Subirachs (1927–2014).

CLUE: PAGE 147 • ANSWER: PAGE 176

Digit Addition

This puzzle is based on a series invented by the Indian mathematician D. R. Kaprekar (1905–1986).

What number comes next?

28, 38, 49, 62, 70, 77, 91 . . .

Some research exists suggesting that humor and puzzles may be related or located in the same area of the brain. Perhaps this is not so surprising after all: Humor makes us laugh out loud, "Ha-ha!" while solving a tough puzzle leads us to exclaim, "Aha!"

Gangster Talk

This type of puzzle always presents a challenge. You've done a few like it, so enjoy it one last time.

> Five gang members were brought in by the police yesterday for questioning. One was suspected of having murdered a rival in another gang. Here's what each one said.
>
> GARY: **Hank did it.**
> WALTER: **Yes, Hank is the killer.**
> JACK: **Hank didn't do it.**
> SAM: **Gary did it.**
> HANK: **Yes, Gary is the killer.**
>
> Four of the suspects lied and only one told the truth. Strangely, the single truth-teller was the murderer. Can you identify the killer?

Martin Gardner once made the following statement, which sums up the subtext of this book perfectly: "The sudden hunch, the creative leap of mind that 'sees' in a flash how to solve a problem in a simple way, is something quite different from general intelligence."

CLUE: PAGE 147 • ANSWER: PAGE 177

A Take on Phillips's Masterpiece

Let's end with a modified (and less complex) version of a master-piece from the pen of British puzzle-maker Hubert Phillips, who we have met several times in this book. The reasoning seems harder than it is, in this case.

Before they are blindfolded, three women are told that each one will have either a red or a blue cross painted on her forehead. When the blindfolds are removed, each woman is then supposed to raise her hand if she sees a red cross and to figure out the color of the cross on her own head. Now, here's what actually happens. The three women are blindfolded and a blue cross is drawn on each of their foreheads. The blindfolds are removed. After looking at each other, the three women do not raise their hands, of course. After a short time, one of the women says, "Our crosses are all blue." How did she figure it out?

Bonus Puzzle

A classic puzzle goes like this:

Each of three houses has three utilities: gas, electricity, and water. Connect each house to all three utilities, so each house has three lines, and each utility also has three lines. In setting this up, you cannot cross lines. You also cannot pass lines through houses or utilities. You cannot share lines either. Can you draw the nine lines required?

The answer to this puzzle is not provided here, since this is an "extra" for you to contemplate, but you can follow up on your own. Make no mistake, it's a tough nut. You will find explanations on various websites, but before you do, try it out yourself.

CLUES
✳

LEVEL 1: SMARTYPANTS

1. If it helps, you can work backward, starting with a common word for "saliva." Also keep in mind that there are four letters involved. Another clue for "stain" is "blemish."

2. How would you describe the legal status of people who are husband and wife?

3. You really don't need any help with this one—all you have to do is consider three coins (other than a nickel) that add up to 45 cents. You may have to try out a few combinations, if you do not immediately see it. That's okay!

4. Think of how you are related to the male person who would have a mother-in-law in your family.

5. Keep in mind that the accountant is an only child. Which of the three people does this condition exclude for that position?

6. Here are two compound parts, in no particular order: *look* and *store*. Does this help?

7. Set each person in a left-to-right order with respect to each other using symbols, such as H = Hannah and G = Gina. For example, if you are told that Frieda beat Hannah, you can show it as: F—H, which means that F comes before H. Be careful though. In this type of puzzle such arrangements may be only temporary, since other people may come in between. But that is part of the fun of solving such puzzles.

8. Think about how each item would be classified scientifically, according to its physical composition.

9. Think of each letter as representing the first letter of a specific word. Also, keep in mind the words are connected in a logical sense, as well.

10. Start by calculating the amount Bill drank with respect to the number of pints that Lucy drank.

11. As indicated in the puzzle statement, you have to assume the worst-case scenario—drawing out two balls of different colors. What happens on your next draw?

12. Like puzzle 11, assume the worst-case scenario—drawing out balls of different colors in a row.

13. The word to be added on would make each word a compound word. For example, the word *hand* can be added to *kerchief* or *maiden* to produce *handkerchief* and *handmaiden*. With some trial and error, you will have to narrow down the choice so that the word will make sense when added to all the given words.

14. Observe how each successive number in the sequence increases with respect to the numbers that come before it.

15. In this case, each number increases by a certain factor. Figure that out and you've solved it.

16. Keep in mind that the word refers to a type of vessel.

17. As with puzzle 16, keep the meaning in mind. The word means something that makes things go around.

18. Let the fish's length be x. How would you represent half of this length? You can then set up an equation easily.

19. What letter in *cold* can be changed to form a legitimate word that will likely lead to *warm*? After deciding on that letter, just follow your word sense.

20. The key trip over is the first one, of course. Imagine what would happen if you started with the wolf, the goat, or the cabbage first. There is only one way to avoid dire consequences.

21. The word to be inserted contains four letters.

22. Consider very carefully what the statement *50 cents more than the eraser* means. Note that the pencil does not cost 50 cents, but this much more than the other item.

23. Call the trains A and B, and draw a sketch of where they are with respect to New York when the two *meet*. Look closely at your sketch.

24. Say the words out loud as you see them laid out, starting with *duck*.

25. Think of the phases of the day—dawn, noon, and twilight— as phases of human life.

LEVEL 2: PRODIGY

26. What state is a dormitory usually in? Neat, messy ... ? What two-word phrase describes a typical dormitory at college?

27. Traditionally, geometrical figures are classified as two-dimensional or three-dimensional. Does this help?

28. The box must be one from which, after you have taken out one tie, you can easily determine the color of the other.

29. Try a simpler version, with four balls. Then try it with five. Do you see any pattern?

30. Always count the number of butts and remember that three of them equal a new cigarette.

31. There are six steps between *iron* and *lead*.

32. Figure out where Claudio might stand in relation to the others.

33. Consider the opposite: How many sisters does the brother have? Does this help?

34. Keep in mind that each successive number increases "exponentially."

35. Look at the letters that are the same, and start with those, trying out logical number substitutions.

36. Again, examine the letters that are the same, and start with those.

37. Consider the meaning of one of the two verbs in the statement.

38. When you have been waiting a while, you might end up saying "It has been a long..."

39. Start by unscrambling the easy ones—WOT and OD.

40. Again, start with the easy ones.

41. Take a sampling and count the 1s in each sample: the numbers from 1 to 9, from 10 to 19, and so on.

42. Start by pouring water from the 5-gallon container.

43. Simplify by turning "father's father's son" into a clearer expression.

44. The two are four-letter words.

45. Try out a few alternatives first.

46. Take out some coins and see how they can be combined into different values.

47. The book is a well-known one by Dr. Seuss.

48. Consider the statements that say, more or less, the same thing, in figuring out if they are likely to be true or false.

49. Translate "Duma" into English, focusing on what it could mean under the circumstances.

50. Like other doublets, make your first change to reflect what's to come.

LEVEL 3: BRAINIAC

51. Don't forget that shoes have an orientation, so a match will involve a right-foot and a left-foot, not just the same color.

52. Where is the snail in relation to the bottom of the well on the 26th day? Take it from there.

53. Recall that Princess Diana died in an automobile accident.

54. Always start the sequencing with the person who comes in right after someone else; then add the rest later.

55. The word *bee* should give it away.

56. Historically, one part of the newly formed word means "outer space."

57. Think of what part of a truck can be altered easily and temporarily.

58. An alternate meaning of the desired word is "just."

59. If you divide something by 3, then you are dividing it into groups. So, for example, 6 socks divided by 3 makes 2 groups of 3 socks each.

60. If any villager is asked, "Are you a truth-teller?" what answer would a truth-teller give? What answer would a liar provide?

61. Each letter is the first in a series of words logically connected to each other.

62. Again, each letter is the first in a series of words that are logically connected.

63. These letters stand for words in a saying.

64. Replace *event* with *date*.

65. Let the brick's weight be x, and take it from there.

66. An opposite for the word meaning "perspicuity" is *obscurity*.

67. As in all sequences, figure out the value by which each successive number increases.

68. Start by eliminating couples with the same color clothing.

69. Carefully inspect the letters in each word.

70. Figure out who lied first.

71. You may have to try out a few possibilities. Just use your basic math sense.

72. Again, you may just have to try out a few possible combinations of signs.

73. The word meaning "paramour" can also mean "darling."

74. The word meaning "caring" can also mean "sensitive."

75. Figure out who the violinist is first, as this is the easiest to determine.

LEVEL 4: MASTERMIND

76. If you are having difficulty envisioning a solution, take two coins (such as two quarters) and revolve one around the other, tracing the movement with a pencil on a sheet of paper below the coins. Then take a look at the tracing.

77. Draw a vertical number line, placing the firefighter at 0 as the mid-point. Then count from there, going up and down the line.

78. As in the other racing puzzles, start by placing someone who comes right after someone else and work from there.

79. With all puzzles of this type, try a few calculations to narrow down the candidate numbers. After some trial and error, a pattern may emerge that will allow you to envision the answer.

80. Again, try a few calculations with prime numbers to narrow down the ones that are potential candidates. Trial and error may help.

81. As before, remember to apply worst-case scenario reasoning.

82. Again, apply worst-case scenario logic.

83. As with other puzzles in this genre, try a few calculations to narrow down the candidate numbers.

84. Here we go again: Try a few calculations to narrow the field of candidate numbers.

85. Yet another example: Try a few calculations to narrow down the candidates.

86. Rephrase "sister's nephew" and "man's son" in different ways. This might help you untangle the connection of these relations.

87. Rephrase "[Mary's] mother's grandson" in your own words. Again, this might help you unravel the relationships.

88. Again, rephrase the confusing statements into your own words. This allows you to envision the relations more concretely.

89. There are six steps between the two words.

90. Set up a chart with all possibilities, and take it from there.

91. There are six steps.

92. The best way to decode a cryptogram is to figure out what word or words are the most likely to occur first in an English sentence, such as articles and demonstratives.

93. First figure out the uphill and downhill rates as two separate speeds.

94. A synonym for the word meaning "shred" is *grind*.

95. A synonym for the word meaning "inventive" is *imaginative*.

96. A synonym for the word meaning "fatherly" is *protective*.

97. Remember that a pair of socks must match not only in color but also in orientation (right-foot and left-foot).

98. Figure out what the two probabilities are together.

99. Rephrase "this man's son" and "my father's son" in your own words. As always, this might help you unravel the relation involved.

100. Set up a timeline or chart for the required weeks, and indicate the days when Jason works and the days when Alicia works. It will then be easy to see when these two line up with each other.

LEVEL 5: GENIUS

101. Think of the various possibilities for placing the 0, since it is a unique number with unique properties.

102. The letter *A* has been replaced here with the number 26.

103. Set up a diagram showing different positions for the trains.

104. For each option, calculate your salary at the end of each year.

105. As in the case of puzzle 49, translate "Ruma" logically into English.

106. Start with the contradictory statements—one of which will be true and the other false.

107. Set up a sequence, moving from left to right, in which the numbers are placed according to the given statements. You may have to try out a few possibilities.

108. Set up a table showing the possible outcomes of the two dice for throwing a 6 or a 7.

109. Make the following substitution: Who am I = What day (today, yesterday, tomorrow, and so on) am I?

110. There are several ways to solve this, but as in puzzle 20, the first trip over is the crucial one. How does the traveler start his journey: with the wolf, the Wolf-Eater, the goat, or the cabbage?

111. There are five steps.

112. There are six steps.

113. There are 10 steps.

114. One of the letters in the word is *I*, which is the roman numeral for 1.

115. Set up the numbering of the houses in categories: houses 1 through 9, houses 10 through 19, and so on.

116. Notice that there is something specific about the shape of the letters.

117. There are 13 steps.

118. Who is the daughter's mother in this case? Rephrase it in your own words.

119. Consider the position of 0, since it has special properties.

120. A synonym for the word meaning "opposition to" is *struggle*.

121. A synonym for the word meaning "inferences" is *conclusions*.

122. The 1s can be digits in a number.

123. As in all these sequences, figure out how the value of each successive number is determined. In this case, look at the actual digits of each number.

124. Determine who the liars are by comparing statements that say the same thing.

125. Enter the mind of the woman who solved it, and reason as she did.

ANSWERS
*

LEVEL 1: SMARTYPANTS

1. *Spot—Spit.* By changing the *o* in *spot* (*stain*) to *i*, you will get *spit* (*saliva*).

2. The rearranged word is *married.* Successfully married people are also admirers of each other, aren't they?

3. *A quarter (25 cents) and two dimes (20 cents).* No further explanation is needed, correct?

4. The man is the girl's *father.* Her father's mother-in-law is his wife's mother, and his wife is the girl's mom.

5. *Bob = engineer, Janet = accountant, Shirley = director.* We know that Bob has a brother, so he is not the accountant, who is an only child. And Shirley earns more than someone else (the engineer, to be exact), so she is also not the accountant, who earns the least. Who does this leave for the accountant position? The only possibility is Janet. We are told that Shirley earns more than the engineer, so she cannot be the engineer. What's left for Shirley then? She isn't the accountant or the engineer, so she must be the director. The final position of engineer is therefore held by Bob based on the process of elimination.

6. (1) *friendship* (friend = companion, ship = sailing vessel),
(2) *bookstore* (book = volume, store = retail establishment),
(3) *greenhouse* (green = leaf color, house = abode),
(4) *lookout* (look = gaze, out = away)

7. *Gina.* Let's represent the outcome as follows: Gina—
Frieda—Hannah. This shows that Frieda beat Hannah but
not Gina, which means that Gina came in ahead of Frieda.

8. *Gasoline* is the odd one out because it is a liquid; the others
are solids.

9. *The letters represent the first letters in the days of the week:*
Monday, Tuesday, Wednesday, Thursday, Friday, Satur-
day, Sunday.

10. *Lisa drank ¾ pint.* If Lucy drank 6 pints, then Bill drank half
of this, which is 3 pints; Arnie drank half of this amount, or
1½ pints; and Lisa drank half of this, or ¾ pint.

11. *Three draws.* Suppose the first ball you draw out is a white
one. If you're lucky the next ball you draw out will also
be white, and it's game over! But you cannot assume this
luck-based scenario. You must, on the contrary, assume the
worst-case scenario; this means that the next ball you draw
out is black. Thus, after two draws, you will have taken
out of the box one white ball and one black ball, under the
worst-case scenario. Obviously, you could have drawn
out a black ball first and a white one second. The end
result would have been the same: one white and one black
ball after two draws irrespective of order. Now, the next
draw—the third one—will be either a white or black ball
(that's what is in the box). Either way, that third ball will

match one of the other two already outside. If it is white, it matches the white ball outside; if it is black, it matches the black ball outside.

12. *Four draws.* The reasoning for the 30-ball, three-color version is the same. You start by assuming the worst-case scenario. What is that? It consists of drawing out three balls of three different colors—white, black, and red, in any order. Now, keep in mind that these three balls of different colors are now outside the box. The fourth ball you draw out—no matter what color—will match one of the three colors outside the box, since it can only be white, black, or red.

13. *Bird: jailbird, birdcage, birdbrain, birdcall, birdsong.*

14. *89.* Each number, starting with the third one, is the sum of the previous two.

15. *39.* Each successive number increases by one more than the last. So, 3 increases by one to produce 4; 4 increases by two (one more) to produce 6; 6 increases by three (one more from the last) to produce 9; and so on.

16. *Kayak.*

17. *Rotator.*

18. *40 feet.* The trout is 20 feet plus half its length. Let's say its length is x, so, its full length is: $x = \frac{1}{2}x + 20$. This says that the total length (x) equals 20 feet plus half the overall length ($\frac{1}{2}x$). Solving for x we get 40.

19. The steps are as follows:
COLD—(1) *cord*—(2) *word*—(3) *ward*—WARM.

20. *Three back-and-forth trips and one final one-way trip over at the end.* The traveler cannot start with the wolf, since that would leave the goat alone with the cabbage, and the goat would eat it. Also, the traveler cannot start with the cabbage, since that would leave the wolf and goat alone, and the wolf would eat the goat. So, practically speaking, the traveler can only start by taking the goat with him on the boat to the other side, leaving the wolf safely with the cabbage on the original side. After dropping off the goat on the other bank, he then rows back alone. This constitutes his first round trip. Back on the original side, he picks up the wolf and rows with it to the other side, leaving the cabbage behind. Upon reaching the other bank he drops off the wolf, but rows back with the goat, so that the wolf cannot eat the goat for lunch. Again, this decision is just common sense, and constitutes the traveler's second round trip. Back on the original side, he leaves the goat there, and this time, takes the cabbage with him on the boat. When he reaches the other bank, he drops off the cabbage, leaving the wolf and cabbage safely together, as he rows back alone. This is his third round trip. He then picks up the goat on the original side and rows across with it. When he reaches the other bank he will have his wolf, goat, and cabbage intact and can finally continue on with his journey.

21. *Life.* The two words created are *wildlife* and *lifestyle*.

22. *The eraser costs $2\frac{1}{2}$ cents.* This means that the pencil costs $52\frac{1}{2}$ cents, which is 50 cents more than the eraser. When you add the two together, the total is 55 cents.

23. *Neither train will be nearer to New York. They will be the same distance from the city.* The trick in this puzzle is the word *meet.* Everything else in the puzzle is useless information.

Obviously, when the two trains meet, neither train will be nearer to New York, even though they are going in different directions.

24. *Duck under the bridge,* with *duck* referring to the verb meaning "lower the head or the body to avoid a blow."

25. The fearless Oedipus answered: *"Humans,* who crawl on all fours as babies (the dawn of life), then walk on two legs as grown-ups (the noontime of life), and finally need a cane in old age (the twilight of life) to get around." Upon hearing the answer, the astonished Sphinx killed itself, and Oedipus entered Thebes as a hero. Ironically, by solving the riddle, the devastating prophecy Oedipus had tried to elude came true—a divination that he would kill his father (a crime he had unwittingly committed on the way to Thebes) and marry his mother, the widowed queen of Thebes.

LEVEL 2: PRODIGY

26. *Dirty room.* It's fascinating how words are connected to each other, isn't it? Maybe the ancients were right after all, believing that anagrams reveal unexpected truths about the world. Bizarre!

27. *Cube* is the odd one out because it is a three-dimensional figure; the others are all two-dimensional.

28. *The actual contents of each box can be determined in just one drawing if that drawing is made from the box labeled BW. All three labels are incorrect. So, the contents of the one labeled BW should actually be either BB or WW. If you draw out a black tie, then its contents are, of course,*

BB; alternatively, if you draw out a white tie, then its contents are *WW*. Either way, you will have identified its contents. The contents of the other two boxes can now be easily determined. It all depends on which color you draw from the box labeled *BW*. Here are the two possibilities:

Scenario 1: The *BW* box is mislabeled. It really contains two black (*BB*) or two white (*WW*) ties. Assume that from the *BW* box you draw out a black tie (*B*). This means that its contents are *BB* (the two black ties). Now consider the *WW* box, which is also mislabeled. It will contain either *BW* or *BB* (not *WW*, of course, as its wrong label claims). It cannot contain *BB*, since the first one does, as just deduced. So, it must contain *BW*. By elimination, the third box, mislabeled *BB*, contains *WW*.

Scenario 2: Again, the *BW* box is mislabeled. It really contains two black (*BB*) or two white (*WW*) ties. Assume this time that you draw out a white tie (*W*). This means that its contents are *WW* (the two white ties). Now consider the *BB* box, which is also mislabeled. It will contain either *BW* or *WW* (not *BB* of course). It cannot contain *WW*, since the first one does, as just deduced. So, it must contain *BW*. By elimination, the third box, mislabeled *WW*, contains *BB*.

29. Take three balls and place them on the left pan. Take three others and place them on the right pan. Put the remaining seventh ball on the table. Now, consider the scale. If the pans balance, then the six balls are all the same weight, meaning that the ball on the table is the culprit ball. However, we can't assume this very fortunate outcome. As in most puzzles of this kind, we must assume a worst-case scenario, since we are asked to weigh the balls no more than twice. So, let's assume that one of the pans goes up, indicating that it contains the culprit ball. Let's eliminate the other three balls. Now, for our second time on the scale, we take the three balls on the pan that went up, put one on the table, and each of the other two on separate pans. There is only one possible outcome left. If the pans balance, then the ball on the table is the culprit ball. If one of the pans goes up, then it contains the culprit ball. Either way, in weighing the balls two times, we are now sure we have identified the ball that weighs less.

30. *40 cigarettes.* Jack smoked the 27 cigarettes he took out from his pocket. Since he smoked only $\frac{2}{3}$ of a cigarette, he therefore would leave a butt equal to $\frac{1}{3}$ of a cigarette. So, for every 3 cigarettes he smoked, he was able to piece together a new cigarette ($\frac{1}{3}$ butt + $\frac{1}{3}$ butt + $\frac{1}{3}$ butt = 1 new cigarette). After smoking the original 27 cigarettes, he was thus able to make 9 new cigarettes. If you stopped here, simply adding 27 (number of cigarettes Jack smoked originally) + 9 (number of new cigarettes made and smoked by Jack) = 36 (total number of cigarettes), you forgot that smoking the 9 new cigarettes also produced butts. In fact, Jack's 9 new pieced-together cigarettes produced 9 new butts of their own. From these 9 butts, Jack was able to make, of course, 3 more cigarettes (3 butts = 1 new cigarette). So, in addition to the 9 new cigarettes Jack made from the original 27, he was also able to make 3 more from those 9 pieced-together ones. But, then, those 3 extra cigarettes produced 3 butts of their own, from which Jack was able to make yet 1 more cigarette. Altogether, therefore, Jack smoked 27 + 9 + 3 + 1 = 40 cigarettes before giving up his bad habit.

31. *Iron*—(1) *icon* (change the *r* in *iron* to *c*)—(2) *coin* (rearrange the letters in *icon*)—(3) *corn* (change the *i* in *coin* to *r*)—(4) *cord* (change the *n* in *corn* to *d*)—(5) *lord* (change the *c* in *cord* to *l*)—(6) *load* (change the *r* in *lord* to *a*)—*Lead* (change the *o* in *load* to *e*). In summary: IRON—*icon*—*coin*—*corn*—*cord*—*lord*—*load*—LEAD. Another possible solution is IRON-icon-coin-loin-loan-lean-LEAD.

32. As for the other puzzles in this genre, the final order reflects all the statements. The only order that works (using the players' initials: A = Armand, C = Claudio, S = Shirley, D = Dina, E = Elgin) is A—S—D—E—C. Translate this into the statements made and you will see that it holds.

33. *Frank has four children*—three daughters and one son, who is, of course, a brother to the three sisters.

34. *729.* The numbers increase as powers of 3: 3^1 (= 3), 3^2 (= 9), 3^3 (= 27), and so on.

35. $T = 3, I = 1, P = 2, A = 5$. TIP + PIT = APA: 312 + 213 = 525. These numerical substitutions are the ones that work mathematically—the solver simply has to do the substitutions to see it.

36. $S = 4, L = 5, O = 0, B = 1, K = 6$: SLOB + BLOL = KOOK: 4501 + 1505 = 6006. These substitutions are the ones that work mathematically, as in number 35. Another possible solution is 4503 + 3505 = 8008.

37. *Only one.* Only one person was *going* to St. Ives—the narrator of the rhyme. If you read carefully, you'll see it never says which direction the kits, cats, sacks, and wives were going. In fact, they were *coming from* St. Ives, moving in the opposite direction from the narrator!

38. *Time.* Time is indeed both long and short, and so on, as Voltaire points out.

39. *Two wrongs do not make a right.*

40. *To err is human; to forgive, divine.*

41. *1.* The digit occurs once in each of the numbers from 11 to 19, and also in every subsequent set—21, 22 . . . ; 31, 32 . . . . Logically, it appears more than any other number.

42. First, Mark fills the 5-gallon container. He pours 3 gallons of it into the 3-gallon container, filling it up; the remaining 2 gallons he then pours into the 10-gallon container. He then fills up the 5-gallon container with water again and adds this to the 10-gallon container, thus producing the required 7 gallons in the 10-gallon container.

43. *Father.* The father's father is Beverley son's grandfather. The son of the grandfather is, of course, the father himself.

44. *Lead–Deal:* To lead is to guide someone (definition 1) and a deal is a bargain (definition 2).

45. *One of each.* Mary has just three pets: She has one dog, one cat, and one rabbit. This works out perfectly, since all her pets except two (3 − 2; that is, 1) are supposed to be dogs. And, indeed, she has one dog. Similarly, she has 3 − 2 (that is, 1) cat, and 3 − 2 (that is, 1) rabbit.

46. *7 dimes and 13 nickels.* First, assume that Alex has an equal number of dimes and nickels in his pocket—10 of each. How much money does that make? Well, 10 dimes add up to $1, and 10 nickels to 50 cents, for a total of $1.50. The total is obviously too high, because Alex has only $1.35 in his pocket. So clearly, fewer dimes are needed in the addition scenario. If a dime is taken away from the 10, then the number of nickels must be increased by 1—because Alex has 20 coins in his pocket. So, 9 dimes and 11 nickels add up to $1.45. This total is still greater than $1.35. So, try reducing the number of dimes in Alex's pocket by 2. This would give him 8 dimes and 12 nickels (adding up to 20 coins), adding up to $1.40. This total is still too high, but we are getting closer to the goal of $1.35. So, let's see what happens when the number of dimes in Alex's pocket is reduced by 3 to a total of 7. This brings him to 7 dimes and 13 nickels, which adds up to $1.35.

47. *The Cat in the Hat.* As you can see, the word *cat* is inserted in the word *hat* after the *h*. As you may know, the story revolves around a tall cat, who wears a red-and-white-striped hat and a red bow tie. It is a hilarious tale loved by children across generations.

48. *Edwin.* The first three statements say the same thing— namely that Earl is the murderer. So, they are either all true or all false. They cannot be true, since there was only one true statement in the set. So, they are all false. We can now see that Edwin's statement is true—Emma did indeed lie, as we just found out. This being the case identifies Edwin as the murderer—the only one who told the truth. Earl's statement is obviously false but changes nothing.

49. *Second individual = Bawi, third individual = Mawi.* The key is to translate "Duma" into English. The translation is "I am a Bawi." Let's break down why this is so. If the individual were a Bawi, he would claim to be one, since Bawis always tell the truth. So, in this case "Duma = I am a Bawi." Now, if he were a member of the dishonest Mawi tribe, would he admit to it? Of course not. So, he would lie and say again "Duma = I am a Bawi," but it would be a lie this time. Either way, "Duma = I am a Bawi." The second individual clearly told the truth, while the third one lied. Finally, it is not possible to determine the tribe to which the first individual belonged.

50. Here are the links: LASS—(1) *mass*—(2) *mast*—(3) *malt*— MALE.

LEVEL 3: BRAINIAC

51. *The answer is 13.* The box contains 24 shoes in total: 6 pairs of black shoes = 12 black shoes; 6 pairs of white shoes = 12 white shoes. Of the 24, half are right-foot-fitting and half are left-foot-fitting. In a worst-case scenario, we might pick all 12 left-foot-fitting shoes (of which 6 are black and 6 are white) or all 12 right-foot-fitting shoes (of which 6 are black and 6 are white). The 13th shoe drawn, however, will match one of these 12.

52. *27 full days, and on the 28th day the snail crawls out.* Since the snail crawls up 3 feet, but slips back 2 feet, its net distance gain at the end of every day is, of course, 1 foot up from the day before. To put it another way, the snail's climbing rate is 1 foot per day. So, on day two, the snail starts at 1 foot from the bottom. On day three, it starts at 2 feet from the bottom, and so on. Now, let's project forward to day 27, where it starts at 26 feet from the bottom and 4 feet from the top. It goes up to 29 feet from the bottom and slides down 2 feet to 27 feet from the bottom. On the 28th day it starts at 27 feet from the bottom, goes up 3 feet, reaching the top, at which point it crawls out of the well. Game over. So, it took 27 full days and nights, and on the 28th day the snail crawled out.

53. *End is a car spin.* It is mind-boggling to contemplate how a name might indeed harbor some prophetic message. Is it just coincidence? You decide.

54. *Thomas.* As for the other puzzles in this genre, the final order reflects all the statements as follows (R = Rashad, M = Mary, J = Jack, W = Walter, T = Thomas): T—M—R—J—W.

55. *Believable.*

56. *Astronaut.*

57. The child suggested deflating the tires, which lowered the height of the truck, allowing the driver to move it through.

58. *Right.* ($tr\underline{i}p$ + $ch\underline{i}n$ + $enou\underline{g}h$ + $si\underline{gh}$ + $\underline{t}emper$ = r + i + g + h + t = $right$.)

59. *59 socks.* Counting a number of socks by 1s, 2s, 3s, and so on is the equivalent of dividing that number into smaller groups of 1 sock, 2 socks, 3 socks, etc. So, to solve this puzzle, you must identify the number of socks between 50 and 60, which, when divided by 3, gives a remainder of 2 (equivalent to saying 2 socks left over), and when divided by 5, gives a remainder of 4. First, divide the numbers between 50 and 60 by 3, identifying those that leave a remainder of 2:

$50 \div 3 = 16$, remainder = 2

$51 \div 3 = 17$, no remainder

$52 \div 3 = 17$, remainder = 1

$53 \div 3 = 17$, remainder = 2

$54 \div 3 = 18$, no remainder

$55 \div 3 = 18$, remainder = 1

$56 \div 3 = 18$, remainder = 2

57 ÷ 3 = 19, no remainder

58 ÷ 3 = 19, remainder = 1

59 ÷ 3 = 19, remainder 2

60 ÷ 3 = 20, no remainder

You can now see that the only candidates between 50 and 60, which, when divided by 3 will leave a remainder of 2, are the numbers 50, 53, 56, and 59. Discard the others, proceeding to determine which of that group (50, 53, 56, or 59) will leave a remainder of 4 when it is divided by 5:

50 ÷ 5 = 10, no remainder

53 ÷ 5 = 10, remainder = 3

56 ÷ 5 = 11, remainder = 1

59 ÷ 5 = 11, remainder = 4

As you can see, 59 socks is the answer. And, in fact, when you count 59 socks 3 at a time, you'll get 2 left over; but when you count them 5 at a time, you'll get 4 left over.

60. *All three men belonged to the liar clan.* Tor's answer, "Yes," implies two possibilities. If Tor is a truth-teller, then his answer implies that Dor is also a truth-teller. If he is a liar, then "Yes" is an expected lie, and thus, so is his response that Dor is a truth-teller. Either way, Tor and Dor belong to the same clan. This is contrary to what Dor himself says. From this we deduce that Dor is a liar. So is Tor, also by deduction. We can now see that Gor lied when he implied that Dor is a truth-teller.

61. *E for eight.* Each letter is the first one in the spelling of consecutive numbers: *O* (one), *T* (two), *T* (three), and so on.

62. *D for December.* Each letter is the first one in the spelling of each month of the year in order of the 12-month calendar: *J* (January), *F* (February), *M* (March), and so on.

63. *W for words.* Each letter is the first one in the spelling of each word in the saying, *A picture is worth a thousand words.*

64. *Independence Day, July 4, 1776.* If you separate the digits as 7-4-1776, you will get the numerical version of the date on which the United States declared its independence from Britain.

65. *3 kilograms.* If the brick weighs x, then $\frac{3}{4}$x stands for $\frac{3}{4}$ its weight. Together with $\frac{3}{4}$ kilograms we get the balance with x kilograms. So, x = $\frac{3}{4}$x + $\frac{3}{4}$, which means that x = 3.

66. *Charity—Clarity.*

67. *10,080.* The first number is multiplied by 2 (2 × 2 = 4), then each successive number is multiplied by a number that is one greater each time: 4 × 3 = 12, 12 × 4 = 48, 48 × 5 = 240, 240 × 6 = 1440, and finally 1440 × 7 = 10,080.

68. *The boy dressed in red dated the girl dressed in blue; the boy dressed in green dated the girl dressed in red; and the boy dressed in blue dated the girl dressed in green.* You are told by one of the boys that no one had a date with a partner dressed in the same color: That is, the boy dressed in red did not have a date with the girl dressed in red; the boy dressed in green did not have a date with the girl dressed in green; and the boy dressed in blue did not have a date with the girl dressed in blue. The boy who made this observation was dressed in red, and he was not dancing with the girl dressed in green, since he made the observation to her and her partner. So, you can safely eliminate the girl in green as a possibility for the boy in red. This means the boy in red dated the girl in blue. Therefore, the boy dressed in green dated the girl dressed in red—since this is the only possibility left. The rest is straightforward.

69. *Stop.* All the words are anagrams of each other.

70. *Gary.* Alex and Tara clearly contradict each other, so one of their statements is true and the other false. Whichever is true, we have now identified that the remaining statements must be false, meaning that Daniela and Gary lied. From this, we can see that Gary, contrary to what he said, is our robber.

71. $34 \div 43 - 6 = 71$.

72. $72 \div 8 = 7 + 2$.

73. *Liver—Lover.*

74. *Sender—Tender.*

75. *Rhonda = violinist, Bernard = drummer, Peter = singer, Selena = pianist.* We can eliminate Peter as the violinist because we know he attended many of the pianist's concerts with the violinist. Selena and Bernard can be eliminated, as well, since they often play with the violinist. So, that leaves Rhonda as the violinist. We are told Peter is not the drummer, and again, he has attended the pianist's concerts, so by deduction Peter is the singer. Then Selena is not the drummer because we are told that the drummer often performs with her. She is also not the violinist (Rhonda is) nor the singer (Peter is). So, by elimination she must be the pianist. This leaves the drummer as the only possibility left for Bernard.

LEVEL 4: MASTERMIND

76. *2 revolutions.* Many people come up with an incorrect solution to this puzzle. Since the circumferences of the two coins are equal, and since the circumference of A is laid out once along that of B, they argue that A must make one revolution about its own center. However, if you actually carry out the instructions of this puzzle with, for example, two quarters, you will find that the outer quarter will make two complete revolutions, not one. The mathematical explanation of this apparent paradox can be found in an analysis of the figure known as a *cycloid*—this is defined as a curve tracing the path traversed by a point on the circumference of a wheel as it rolls without slipping upon a straight line. The cycloid was studied and named by Italian physicist and astronomer Galileo in 1599.

77. *25 rungs.* At the beginning, we do not know what rung the firefighter is standing on, except that it is the "middle" rung. So, let's label her starting rung as 0, that is, consider the middle rung to be analogous to the 0, or ground level, of a building. Each rung above and below 0 can then be compared to a level above or below this ground level. Obviously, since it is the middle rung, there will be as many rungs above it as there are below it. You are first told that the firefighter went up 3 rungs from the 0 rung. You are then informed that she stepped down 5 rungs. So, from rung 3 above 0, she went down 5 rungs, ending up at 2 rungs below the 0 point (which you can also represent as -2). Next, the puzzle tells you that the firefighter climbed up 7 rungs (from rung -2). So, she started from rung -2 and went up 7 rungs from there. This means she ended up at rung 5 above the starting point or 0 rung. Finally, the puzzle tells you that the firefighter climbed up another 7 rungs (from rung 5 above) to the roof. This means that she continued climbing (from rung 5 above), moving up another 7 rungs, to rung 12 beyond her starting point. Rung 12 above is the top part of the ladder, because from that rung the firefighter stepped onto the roof. Now, let's complete the ladder. You know that it has 12 rungs above the 0 rung. Since the 0 rung is the middle rung, a complete ladder will, of course, also have 12 rungs below the zero rung. Therefore, the ladder consists of 12 rungs above the 0 rung, 12 below it, and the 0 rung itself. This makes, of course, 25 rungs in total.

78. *Jerry.* As with the other puzzles in this genre (7, 32, 54), this final order reflects all the statements as follows (J = Jerry, B = Bob, P = Paula, S = Sarah, T = Tim, L = Lorraine): J—B—P—S—T—L.

79. *33.* $33 \times 3 = 99 \div 9 = 11$.

80. *17.* $17 + 2 = 19; 17 \times 3 = 51 - 17 = 34$.

81. *Six draws.* The reasoning is always the same. Assuming the worst-case scenario, you will draw out five balls of different colors—a white, a black, a green, a blue, and a yellow, in no particular order. Now, whichever ball you draw out next (remember that there are still balls of each color left inside in the box), it will match one of the five that have been drawn out. This is the sixth draw. If it is a white ball (from inside the box), it matches the white ball outside; if it is a black ball, it matches the black ball outside; and so on.

82. *Eight draws.* Nothing changes in the reasoning. Assuming the worst-case scenario, you will draw out seven balls of different colors—a white, a black, a green, a blue, a yellow, a brown, and the single red one, in no particular order. Now, whichever ball you draw out next (remember that there are still balls of each color in the box except for red), it will match one of the balls that have already been drawn out, except, of course, the red one—white, black, green, blue, yellow, or brown. That constitutes the eighth draw. If you draw a white ball (from inside the box), it matches the white ball outside; if it is a black ball, it matches the black ball outside; and so on.

83. *123.* 123 × 2 = 246 − 1 = 245.

84. *13.* 13 × 4 = 52. Note the sum of the digits in "13" is "4."

85. *3.* The steps break down as follows:

3 + 4 = 7 ("Add me to the next number above me.")

7 x 3 = 21 ("Multiply the result by me . . .")

21 + 3 = 24 (". . . and add me again.")

2 + 4 = 6 ("Add the digits in the result.")

6 ÷ 2 = 3 (Divide in half for the final answer.)

86. *Wife.* The woman has a sister, since she uses the expression "my sister's." Her sister's nephew in this case is obviously the woman's son. Think of it this way. Suppose you have a sister and you have a son. Who is the son in relation to your sister? Her nephew, of course. The puzzle also says that her nephew is the son of the man in the photo. Conclusion? The woman is looking at a photo of her husband.

87. *Aunt.* Mary's mother is also her sister's mother, needless to say. So, the mother's grandson can only be her sister's son, since Mary has no children. This means that the grandson is her nephew and she is his aunt.

88. *Niece.* The niece of your mother is your cousin. The mother of your cousin is your aunt (even if only by marriage).

89. FOUR—(1) *foul*—(2) *fool*—(3) *foot*—(4) *fort*—(5) *fore*— (6) *fire*—FIVE.

90. *Two out of three.* We let B and W-1 stand respectively for the black and white marbles that might be inside the bag at the start, and W-2 for the white marble added to the bag. Removing a white marble from the bag results in three equally likely combinations of two marbles, one inside and one outside the bag:

INSIDE THE BAG	OUTSIDE THE BAG
(1) W-1	W-2
(2) W-2	W-1
(3) B	W-2

(1) In combination 1, the white marble drawn out is the one added to the bag (W-2), and the white marble inside it (W-1) is the piece originally there.

(2) Combination 2 is the converse of (1): the white marble drawn out is the one originally in the bag (W-1), while the white marble inside it (W-2) is the one that was added.

(3) In combination 3, the white marble drawn out is the one added to the bag (W-2), since there was no white marble originally inside it. The marble that remains in the bag is a black one (B).

In two of the three cases, Carroll observes, a white marble remains in the bag. So, the chance of drawing a white marble on the second draw is two out of three.

91. WHEAT—(1) *cheat*—(2) *cheap*—(3) *cheep*—(4) *creep*— (5) *creed*—(6) *breed*—BREAD.

92. *There was never yet an uninteresting life.* Each letter has been replaced with the second one before it in the alphabet sequence: T has been replaced by R, H by F, E by C, and so on.

93. *3 miles per hour.* This classic puzzle has a tendency to mislead solvers, probably because they assume that the average speed is simply calculated by combining the hiker's two rates (2 mph and 6 mph = 8 mph) and then dividing by 2. They invariably come up with the incorrect answer of 4 mph. Let's calculate the time Sarah took to hike uphill. For the sake of convenience, assume that the distance uphill (and downhill, of course) is 1 mile. You can actually use any other distance, and the reasoning and result will be the same. So, in this case, you are told that her rate uphill is 2 miles per hour. Remember that Distance = Rate × Time. Therefore, her *time up* is: 1 (mile) = 2 (miles per hour) × *time up*, which is $\frac{1}{2}$. Thus, it would take Sarah $\frac{1}{2}$ hour to climb a 1-mile hill. Now, given that her rate downhill is 6 miles per hour, let's calculate her *time down*: 1 (mile) = 6 (miles per hour) × *time down*, which is $\frac{1}{6}$. So, it would take Sarah $\frac{1}{6}$ hour to descend a 1-mile hill. Now, her total time for the entire trip is, of course, *time up* + *time down*, or $\frac{1}{2} + \frac{1}{6} = \frac{2}{3}$ hour. The total distance she covered is 2 miles—1 mile up + 1 mile down. Therefore, her overall rate is: 2 (miles) = *rate* × $\frac{2}{3}$ (hours), which is 3 miles per hour. As you can see, Sarah's overall, or average, rate of speed, which has been calculated by taking into account the overall distance she covered (2 miles) and the overall time she took ($\frac{2}{3}$ hour), is 3 miles per hour, not 4 miles per hour!

94. *Grate—Great.*

95. *Creative—Reactive.*

96. *Paternal—Parental.*

97. *31 draws.* Again, we cannot count on luck and must assume the worst-case scenario. In this case, you might get all left-foot or right-foot socks first. So, you will have drawn out 30 socks in total. At this point, the next sock you draw out (the 31st) will match one of these, since only right-foot or left-foot socks are inside and one of these will match in color one of the 30 outside the box.

98. $\frac{2}{13}$. The probability of drawing an ace is $\frac{1}{13}$ and the probability of drawing a king is also $\frac{1}{13}$. The probability of drawing either one or the other is the sum of these two: $\frac{1}{13} + \frac{1}{13} = \frac{2}{13}$.

99. *The boy's father.* The boy is an only child, because he says that he has no brothers or sisters. So, when he refers to "my father's son" he is referring to himself, the only son of his father. Therefore, the son of the man in the photo ("this man's son") is himself ("my father's son"). This also means, of course, that the man in the photo is his father.

100. *Saturday, October 13.* Set up a time chart for the week of October 1, showing that Jason (= J) works every third day and Alicia (= A) on Saturday:

WEEK OF OCTOBER 1

CLERK	MON OCT. 1	TUES OCT. 2	WED OCT. 3	THURS OCT. 4	FRI OCT. 5	SAT OCT. 6	SUN OCT. 7
Jason	J			J			J
Alicia						A	

Since they will apparently not be working together during that week, set up a chart for the week after:

WEEK OF OCTOBER 8

CLERK	MON OCT. 8	TUES OCT. 9	WED OCT. 10	THURS OCT. 11	FRI OCT. 12	SAT OCT. 13	SUN OCT. 14
Jason			J			J	
Alicia						A	

This chart shows that the two will be working together on Saturday, October 13.

LEVEL 5: GENIUS

101. $(22 - 12) - 10 = 0$.

102. *Gossip is the opiate of the oppressed.* The code used for this one is a hard one to crack. Each letter has been replaced by the digits in reverse (backward) order. For example, in the usual order Z would be replaced by 26, since it is the 26th letter of the alphabet; but with this reverse code it is replaced by 1; B would be replaced by 2 in the usual order; now it is replaced by 24, and so on. The complete code (including letters not used in the plaintext) is as follows: $A = 26, B = 25, C = 24, D = 23, E = 22, F = 21, G = 20, H = 19, I = 18, J = 17, K = 16, L = 15, M = 14, N = 13, O = 12, P = 11, Q = 10, R = 9, S = 8, T = 7, U = 6, V = 5, W = 4, X = 3, Y = 2, Z = 1$.

103. *10 trains.* Let's say that you get on a train at the New York station at 12:00 noon. It could be at any other time of course; the reasoning will be the same. As mentioned in the puzzle, five hours later, at 5:00 PM, your train arrives at the Washington station. Now, you must envision the relative positions of the trains on their way from Washington to New York during those five hours. Keep in mind that the Washington-to-New York trains leave on the half-hour. At the Washington station at 5:00 PM, there is a train ready to leave for New York. Call it A. Obviously, the train that had left the Washington station a half hour earlier, at 4:30 PM, will find itself a certain distance from the Washington station when A is about to leave. Call that train B. You can now complete the diagram showing the relative locations of all the trains leaving the Washington station, bound for New York, between 12:00 PM and 5:00 PM as follows:

NEW YORK									WASHINGTON	
K	**J**	**I**	**H**	**G**	**F**	**E**	**D**	**C**	**B**	**A**
12:00	12:30	1:00	1:30	2:00	2:30	3:00	3:30	4:00	4:30	5:00

Now, when you left the New York station at 12:00 noon, you obviously missed passing the 12:00 K-train that had come from Washington, because it was in the station when your train was leaving. But, as you can see from the diagram above, you passed all the others—the 12:30 J-train (that is, the train that left Washington for New York at 12:30), the 1:00 I-train, the 1:30 H-train, the 2:00 G-train, the 2:30 F-train, the 3:00 E-train, the 3:30 D-train, the 4:00 C-train, the 4:30 B-train, and the 5:00 A-train. That makes 10 trains in all.

104. *Option 2.* After the first year with Option 1, you would receive just the $4,000. With Option 2 on the other hand, you would receive $2,000 after the first six months, but then you would get an increase of $200. So, for the last six months of that year, you would receive $2,200 dollars. Adding the two semesters up, you would get $4,200 at the end of the first year, whereas with Option 1 you would get just the $4,000. Now, what income do both options generate after year two? Well, with Option 1 you would receive an increase of $800 for the year. So, you would end up earning $4,800. But with Option 2, you would earn $2,400 the first semester—the $2,200 you would have started off with at the beginning of the year (i.e. salary from the previous semester) plus the $200 raise you would have gotten for that semester. Then, in the last six months you would get another increase of $200 on top of this new salary: That is, you would earn another $2,600 ($2,400 + $200). Adding up the two semesters, you would receive $5,000 at the end of the second year. If you continue calculating the incomes generated by the two options in this way, you would see that Option 2 generates more income in the long run, and is therefore the better option.

105. *Man = truth-teller, Woman = liar.* As in puzzle 49, let's translate "Ruma" into English by logical deduction. If the woman were a truth-teller, she would admit it, so her response, "Ruma," would mean "Yes." If she were a liar, she would not admit it, and her response, "Ruma," would still mean "Yes," but would be a lie. The man clearly told the truth by saying that she said "Yes." Therefore, he is a truth-teller. He tells Dr. Brown that his partner is a liar, which means that she is indeed a member of the liar clan.

106. *Friday.* Barb and Charlene give contradictory statements—namely that today is Saturday (Barb) and that today is not Saturday (Charlene). So, one statement is true and the other false. With this deduction, we have located who gave the only true statement—either Barb or Charlene. This means that the other statements were all false. Fanny's statement implies that today is Thursday, since she says that tomorrow is Friday. That being false, we can eliminate Thursday.

Emma's statement implies that today is Tuesday, since she says that tomorrow is Wednesday. That being false, we can now also eliminate Tuesday. Dina's statement implies that today is Sunday, since she states that the day after tomorrow (Monday) is Tuesday. So, we can eliminate Sunday, too. Alma's statement implies that today is Saturday, since she says that yesterday was Friday. That being a lie, as well, we can also eliminate Saturday. We can now see that Barb lied, claiming that today is Saturday, so Charlene told the truth. She allows us to eliminate Monday and Wednesday from the list, as we can now see from her statement. This leaves Friday as the only possibility.

107. *Runner 2.* Number 1 and Number 5 came in one after the other. Number 1 was not the winner, because he or she did not end up in the spot corresponding to his number. So, he or she could have ended up in second, third, or fourth, followed right after by Number 5. Here are the three possibilities:

(1) Number 1 (2nd)—Number 5 (3rd)

(2) Number 1 (3rd)—Number 5 (4th)

(3) Number 1 (3rd)—Number 5 (5th)

Since Number 5 did not end up fifth, we can discard possibility 3. Now, we are told that Number 3 came in right after 5. So, let's consider the possibilities now:

(1) Number 1 (2nd)—Number 5 (3rd)—Number 3 (4th)

(2) Number 1 (3rd)—Number 5 (4th)—Number 3 (5th)

Number 4 did not win the race, nor did he or she end up in the fourth spot. This leaves only the fifth spot for Number 4. No other positioning would work for him or her, given the consecutive positioning of Numbers 1, 5, and 3. This leaves Number 2 as the winner. The final order was therefore: Number 2 (1st), Number 1 (2nd), Number 5 (3rd), Number 3 (4th), Number 4 (5th).

108. $^{11}\!/_{36}$. We start by listing the number of ways to throw either a 6 or a 7:

OUTCOME: 6		OUTCOME: 7	
FIRST DIE	SECOND DIE	FIRST DIE	SECOND DIE
1	5	1	6
2	4	2	5
3	3	3	4
4	2	4	3
5	1	5	2
		6	1

There are 36 possible throws of two dice, because each of the 6 faces of the first die is matched with any of the 6 faces of the second one. Of these 36 possible throws, 11 produce either a 6 or a 7 (as the table above shows). Therefore, the probability of throwing either a 6 or a 7 is $\frac{11}{36}$.

109. *Today.* Before its "birth," *today* does indeed have a different name—*tomorrow*. And when it is "laid within the tomb," that is, when it is over, it assumes a new name—*yesterday*. Finally, though it lasts only one day, it changes its name three days in a row ("three days together"): from *yesterday*, to *today*, to *tomorrow*.

110. There are several ways to solve the Wolf-Eater puzzle, all consisting of four round trips (nine individual trips in total: four round trips and a final one over). Here's one. The traveler must start by taking the wolf with him to the other side, leaving the Wolf-Eater with the goat and cabbage on the original side. The Wolf-Eater's presence ensures that the goat will not eat the cabbage. Upon reaching the other bank, the traveler drops off the wolf and rows back alone. This is his first round trip. Back on the original side, he picks up the cabbage, leaving the Wolf-Eater and goat together, and rows with it to the other side. Once there, he leaves the cabbage safely with the wolf and then rows back alone. This is his second round trip. On the original side, he picks up the Wolf-Eater, leaving the goat there alone, rowing with the monster to the other bank. There he drops off the Wolf-Eater, and picks up the wolf for his trip back (so the Wolf-Eater will not eat the wolf), leaving the Wolf-Eater alone with the cabbage. This is his third round trip. Upon reaching the original side, the traveler drops off the wolf, picking up the goat. Once he reaches the other side, he leaves the goat safely with the cabbage and Wolf-Eater, who are already there. (As mentioned earlier, the Wolf-Eater's presence ensures that the goat will not eat the cabbage.) He rows back alone, for the completion of his fourth round trip. Back on the original side, the traveler picks up the wolf, and rows back with it to the other side. He gets off the boat with the wolf, and continues his journey with all four.

111. FLOUR—(1) *floor*—(2) *flood*—(3) *blood*—(4) *brood*—(5) *broad*—BREAD.

112. BLACK—(1) *blank*—(2) *blink*—(3) *clink*—(4) *chink*—(5) *chine*—(6) *whine*—WHITE.

113. RIVER—(1) *rover*—(2) *cover*—(3) *coves*—(4) *cores*—(5) *corns*—(6) *coins*—(7) *chins*—(8) *shins*—(9) *shine*—(10) *shone*—SHORE.

114. *1009.* In Roman numerals, the number is MIX.

115. 15. Houses 1–9 = 1 is used once. Houses 10–19 = 1 is used 11 times (10, 11 [twice], 12, 13, 14, 15, 16, 17, 18, 19). Houses 20–29 = 1 is used once (21). Houses 30–39 = 1 is used once (31). Houses 40–49 = 1 is used once (41). House 50 = not used. TOTAL: 1 + 11 + 1 + 1 + 1 = 15.

116. *X, Y.* All these letters do not change when reflected in a mirror. The letters are listed alphabetically.

117. WINTER—(1) *winner*—(2) *wanner*—(3) *wander*—(4) *warder*—(5) *harder*—(6) *harper*—(7) *hamper*—(8) *damper*—(9) *damped*—(10) *dammed*—(11) *dimmed*—(12) *dimmer*—(13) *simmer*—SUMMER.

118. *The woman herself.* Her daughter's mother is, of course, herself.

119. 17 + 13 – 30 = 0.

120. *Resistance—Ancestries.*

121. *Deductions—Discounted.*

122. 11 + 11 + 1 + 1 = 24.

123. *101.* First, you add the digits in a number, starting with 28. The sum of its digits is 2 + 8 = 10. Now, you add the result of 10 to 28 to get the next number in the sequence, namely 38. Now, add the digits in 38, 3 + 8 = 11. Add this result to 38— 38 + 11 = 49 to get the next number in the sequence. And so on.

124. *Jack.* Gary and Walter say the same thing—that Hank is the killer—so their statements are either both true or both false. They cannot be true because there was only one true statement in the set, so they are false. Similarly, Sam and Hank say the same thing—that Gary is the killer. Again, both statements cannot be true since there was only one true statement in the set. So, they are both false. We have now identified the four liars, who are all innocent: Gary, Walter, Sam, and Hank. This leaves Jack as the only truth-teller and thus the killer. As we can see, he did indeed tell the truth—Hank didn't do it—but his statement changes nothing.

125. If any woman would have raised her hand, it would have meant that woman saw at least one red cross on a forehead. This did not happen. So, one of the astute women figured this out.

FURTHER READING

*

There are a number of books that I consider to be the "classics." I simply list them here alphabetically—not by rank or subject matter. Of course, you may want to add others to your personal list.

Alcuin. *Problems to Train the Young*, around 800 CE. See John Hadley and David Singmaster, "Problems to Sharpen the Young." *The Mathematical Gazette* 76 (1992): 102-126.

This contains more than 50 puzzles, including the river-crossing examples you have solved in this book. None of the puzzles require any expert knowledge to solve them. Alcuin wrote it in an era when there was little or no interest in mathematics in Europe. Originally written in Latin, it was clearly intended as an amusement for the well-educated.

Dmitri A. Borgman. *Language on Vacation*, New York: Scribner, 1965.

Without question, this is one of the greatest books of word puzzles and games.

Claude-Gaspard. Bachet de Mézirac, *Pleasant Problems*. Lyon: Gauthier-Villars, 1612.

This is the first book to organize puzzles in mathematics according to type—weighing, measuring, and so on. They are the prototypes for many puzzle anthologies today.

Lewis Carroll. *Pillow Problems and a Tangled Tale.* New York: Dover, 1885.

This may be the number one puzzle book of all time. It contains 72 mathematical posers ranging from those that can be solved by simple mathematical know-how to those that require more advanced or even expert knowledge. The puzzles were originally printed as a monthly magazine serial, and many readers sent in solutions to those posed in it. A number of these puzzles have been included in this book.

Henry E. Dudeney, *The Canterbury Puzzles.* New York: Dover, Originally 1907.

Published in 1907, this is Dudeney's masterpiece, in which he designs his 114 puzzles on characters from Chaucer's Tales. The book is suitable for young enthusiasts, mathematicians, and veteran puzzlers alike. It is, overall, quite challenging.

Henry E. Dudeney, *536 Puzzles and Curious Problems,* edited by Martin Gardner. New York: Dover, 2007.

For two decades, Dudeney wrote a puzzle column, "Perplexities," for *The Strand Magazine.* Martin Gardner hailed Dudeney as "England's greatest maker of puzzles," unsurpassed in the challenge his inventions continue to provide.

Henry E. Dudeney. *Amusements in Mathematics,* New York: Dover, 1958.

These 430 puzzles, problems, paradoxes, and brain teasers are particularly inspired. They span all genres—numbers, unicursal and route problems, counter puzzles, speed problems, measuring, weighing, packing, clock puzzles, and many more. Chessboard problems, involving the dissection of the board or the placement or movement of pieces, age and kinship problems are also included here.

Martin Gardner. *Aha!* New York: Scientific American. 1978.
For 25 years, Gardner wrote "Mathematical Games and Recreations," a monthly column for *Scientific American* magazine. These columns have inspired hundreds of thousands to delve into his many puzzle books—all of them outstanding. This one is especially ingenious since it revolves around the "Aha Effect" produced by puzzles of all kinds.

Martin Gardner. *Gotcha: Paradoxes to Puzzle and Delight!* San Francisco: Freeman, 1982.
This is a companion volume to the previous one. In it, Gardner takes us through the many pitfalls that puzzles lay for us.

Sam Loyd. *Mathematical Puzzles of Sam Loyd.* Ed. by Martin Gardner. New York: Dover, 1959.
Compiled by another great puzzlist, Martin Gardner, this collection by Sam Loyd, one of the leading puzzlists of modern history, contains some of the most challenging puzzles for mathematicians and general public alike. Many are now classic in the field.

Sam Loyd. *Cyclopedia of 5000 Puzzles, Tricks, and Conundrums with Answers.* New York: Dover. Originally 1914.
This book was compiled by Sam Loyd's son after his father's death. Perhaps the most complete volume of all of his puzzles, it is considered the most magnificent and stimulating collection of puzzles ever assembled.

Raymond M. Smullyan. *What Is the Name of This Book? The Riddle of Dracula and Other Logical Puzzles.* New York: Dover, 2011.
Raymond Smullyan (1919–2017) was a mathematician, logician, magician, and creator of extraordinary puzzles. This is one of the most brilliant books in logic puzzles. These puzzles delve into Gödel's undecidability theorem—a celebrated and important one for the foundations of mathematics—as well as some of the deepest paradoxes of logic and set theory.

INDEX
✳

ABOUT THE AUTHOR
✳

MARCEL DANESI teaches a course on the history of puzzles and their meaning to human life at Victoria College of the University of Toronto. He has been a professor of anthropology at the University of Toronto since 1974. Danesi has written puzzles for *Reader's Digest,* the *Toronto Star,* and *Prevention Magazine* and maintains a puzzle blog for *Psychology Today* in which he discusses the significance of different types of puzzles. He has also published several best-selling puzzle books, such as *The Total Brain Workout* and *The Complete Brain Workout.*